European Foreign Policy in a Decarbonising World

Contributing to the emerging literature on the geopolitical and foreign policy implications of decarbonisation and energy transition processes, this book sheds light on the future of the European Union's (EU) external relations under decarbonisation.

Under the Paris Agreement on climate change, adopted in 2015, governments are committed to phasing out the emissions of carbon dioxide and other greenhouse gases over the coming decades. This book addresses the many questions around this process of decarbonisation through detailed analyses of EU external relations with six fossil-fuel exporting countries: Nigeria, Indonesia, Azerbaijan, Colombia, Qatar, and Canada. The authors systematically examine the six countries' varying dependence on fossil fuels, the broader political and security context, current relations with the EU, and the potential for developing these towards decarbonisation. In doing so, they put forward a series of findings that should hold across varying circumstances and provide a steppingstone to enrich and inspire further research on foreign policy, external relations, and international relations under decarbonisation. The book also makes an important contribution to understanding the external implications of the 2019 European Green Deal.

This volume will be of great interest to students and scholars of European environmental and climate policy, climate diplomacy, energy policy, foreign policy, and climate/energy geopolitics.

Sebastian Oberthür is Professor for Environment and Sustainable Development at the Vrije Universiteit Brussel (VUB) and Professor of Environmental Policy and Law at the University of Eastern Finland (UEF).

Dennis Tänzler is Director and Head of Programme Climate Policy at adelphi, Germany.

Emily Wright works as a Manager for adelphi, Germany, in the field of climate policy and diplomacy.

Gauri Khandekar is Researcher at the Centre for Environment, Economy and Energy at the Brussels School of Governance, Vrije Universiteit Brussel (VUB).

Routledge Studies in Environmental Policy

Guanxi and Local Green Development in China
The Role of Entrepreneurs and Local Leaders
Chunhong Sheng

Environmental Policy in India
Edited by Natalia Ciecierska-Holmes, Kirsten Jörgensen, Lana Ollier and D. Raghunandan

Mainstreaming Solar Energy in Small, Tropical Islands
Cultural and Policy Implications
Kiron C. Neale

EU Environmental Governance
Current and Future Challenges
Edited by Amandine Orsini and Elena Kavvatha

The European Union and Global Environmental Protection
Transforming Influence into Action
Edited by Mar Campins Eritja

Environmental Policy and Air Pollution in China
Governance and Strategy
Yuan Xu

Climate Change Law and Policy in the Middle East and North Africa Region
Edited by Damilola S. Olawuyi

European Foreign Policy in a Decarbonising World
Challenges and Opportunities
Sebastian Oberthür, Dennis Tänzler, and Emily Wright, with Gauri Khandekar

For more information about this series, please visit: *www.routledge.com/Routledge-Studies-in-Environmental-Policy/book-series/RSEP*

European Foreign Policy in a Decarbonising World

Challenges and Opportunities

Sebastian Oberthür, Dennis Tänzler, and Emily Wright

With Gauri Khandekar

Routledge
Taylor & Francis Group
LONDON AND NEW YORK

earthscan
from Routledge

First published 2022
by Routledge
4 Park Square, Milton Park, Abingdon, Oxon OX14 4RN

and by Routledge
605 Third Avenue, New York, NY 10158

Routledge is an imprint of the Taylor & Francis Group, an informa business

British Library Cataloguing-in-Publication Data
A catalogue record for this book is available from the British Library

Library of Congress Cataloging-in-Publication Data
Names: Oberthür, Sebastian, author.
Title: European foreign policy in a decarbonising world : challenges and opportunities / Sebastian Oberthür, Dennis Tänzler, and Emily Wright with Gauri Khandekar.
Description: Abingdon, Oxon ; New York, NY : Routledge, 2022. |
Series: Roultedge studies in environmental policy | Includes bibliographical references and index.
Subjects: LCSH: European Union countries—Foreign relations—21st century. | Environmental policy—European Union countries. | Carbon dioxide mitigation—Political aspects. | Petroleum industry and trade—Political aspects. | Gas industry—Political aspects.
Classification: LCC JZ1570 .O34 2022 (print) | LCC JZ1570 (ebook) | DDC 327.4—dc23
LC record available at https://lccn.loc.gov/2021042217
LC ebook record available at https://lccn.loc.gov/2021042218

ISBN: 978-1-032-01131-8 (hbk)
ISBN: 978-1-032-02352-6 (pbk)
ISBN: 978-1-003-18303-7 (ebk)

DOI: 10.4324/9781003183037

Typeset in Perpetua
by codeMantra

Contents

Illustrations

Figures

Table

Preface and acknowledgements

With the publication of this book, a journey of several years reaches its culmination point. More than five years ago, we discussed the first ideas for investigating broader EU external/bilateral relations under the then still new notion of decarbonisation, since we felt that the literature had yet to more fully grasp the full breadth of this development. This first led to a study for the German Federal Foreign Office on 'The Geopolitics of Decarbonisation: Reshaping European Foreign Relations', which was conducted between 2017 and 2019 and published in 2020. This study formed the basis of this book volume for which we considerably expanded and deepened the discussion in the introductory and concluding chapters, including in relation to relevant academic literatures and discussions. We also undertook to update the six country case studies, including reflections on the still unfolding repercussions of the Covid-19 pandemic on the countries' economic and social situation, which hit just as we finalised the original study. It very much brought to the fore a point that remains valid, namely that the study's subject remains volatile and dynamic and as such this volume can only present a first attempt at an analysis of the various repercussions of decarbonisation on foreign policy as the world advances towards realising the climate transition. We hope that our study can serve as a source of inspiration and a basis for much-needed further work on analysing (European) foreign policy in a decarbonising world.

Whereas Sebastian Oberthür has taken the overall lead in editing this book publication, Dennis Tänzler and Emily Wright have taken the lead on or contributed to all chapters. Gauri Khandekar has taken a co-leading role in the case study on Canada (Chapter 7) and has contributed to the case studies on Azerbaijan and Qatar (Chapters 4 and 6).

We owe thanks to various institutions and individuals. First of all, we are deeply indebted to the German Federal Foreign Office for their support of the original study as well as for turning the study into a book publication as part of the joint cooperation on climate diplomacy. We also wish to acknowledge the sponsorship of the Jean Monnet Network 'Governing the EU's Climate and Energy Transition in Turbulent Times' (GOVTRAN – www.govtran.eu) with the support of the Erasmus+ Programme of the European Union. Furthermore, Angela Van Dijk, Hannah Sofie Forst, and Laura Iozzelli deserve our thanks for their invaluable research assistance at various stages of the project. We hope that the readers will find the overall result of value for advancing our understanding of foreign policy and international relations in a decarbonising world.

Sebastian Oberthür, Dennis Tänzler, and
Emily Wright, with Gauri Khandekar
Brussels and Berlin

Abbreviations

ASEAN	Association of Southeast Asian Nations
CCS	Carbon Capture and Storage
CEPA	Comprehensive Economic Partnership Agreement (Indonesia)
CETA	Comprehensive Economic and Trade Agreement (Canada)
CO_2	Carbon dioxide
ECOWAS	Economic Community of West African States
EEAS	European External Action Service
EIA	Energy Information Administration (US)
EIB	European Investment Bank
ELN	National Liberation Army (Colombia)
EPA	Economic Partnership Agreement (Nigeria)
ETS	Emissions Trading System
EU	European Union
FARC	Revolutionary Armed Forces of Colombia
FDI	Foreign Direct Investment
GCC	Gulf Cooperation Council
GDP	Gross Domestic Product
GHG	Greenhouse gas
IEA	International Energy Agency
IMF	International Monetary Fund
IRENA	International Renewable Energy Agency
LNG	Liquefied natural gas
LULUCF	Land use, land use change, and forestry
ND-GAIN	Notre Dame Global Adaptation Initiative
NDC	Nationally Determined Contribution
NNPC	Nigerian National Petroleum Corporation
OECD	Organisation for Economic Co-operation and Development
OPEC	Organization of the Petroleum Exporting Countries
PLN	Perusahaan Listrik Negara (Indonesian state-owned power utility)
REDD+	Reducing Emissions from Deforestation and Forest Degradation
SISCLIMA	National Climate Change System (Colombia)
SOCAR	State Oil Company of Azerbaijan
SOFAZ	State Oil Fund of the Republic of Azerbaijan

UNDESA	United Nations Department of Economic and Social Affairs
UNFCCC	United Nations Framework Convention on Climate Change
US	United States
WTO	World Trade Organization

1 Introduction

The European Union, decarbonisation, and foreign policy

Introduction

The European Union (EU) and the world community are embarking on a fundamental societal and economic transition away from fossil fuels and towards full decarbonisation (the 'climate transition'). Under the 2015 Paris Agreement, governments agreed to phase out emissions of CO_2 and other greenhouse gases (GHGs) over the coming decades in order to hold global temperature rise to well below 2°C and to pursue efforts to limit it to 1.5°C above pre-industrial levels (Klein et al. 2017). To this end, a growing number of countries have committed to phasing out GHG emissions by 2050 or 2060, including the EU, the United States, China, and others (see https://eciu.net/netzerotracker, last visited on 24 June 2021). The EU has been at the forefront of climate action by international comparison and has generally been considered an international leader on climate change (Oberthür and Dupont 2021). To make the EU sustainable and achieve climate neutrality by 2050, the European Commission launched the European Green Deal in 2019 (European Commission 2019). In this context, the EU has committed to reducing its GHG emissions by 55 percent by 2030 and to achieving climate neutrality (i.e., net zero GHG emissions) by 2050. At the time of writing, further implementing legislation to this end is forthcoming (see the next section).

Transcending climate policy as such, the challenge of the climate transition implies far-reaching adaptations of a wide range of other policies, including foreign policy and external relations.[1] The need for further alignment of other sectoral policies with climate policy arises from the crosscutting nature of the climate challenge. Although climate policy integration – the integration of climate policy objectives into other policies – has significantly advanced in the EU (Dupont 2016), the need for further progress has been acknowledged and highlighted in the European Green Deal that aims to ensure other EU policies support the climate transition. Part of the agenda of climate policy integration concerns EU external policies that need to be further developed so as to build sustainable, fruitful international partnerships under decarbonisation and in support of the climate transition (European Commission 2019: esp. 15–22).

Research on the integration of climate policy into EU foreign policy at large is at an early stage. As further discussed below, several literatures possess relevance for the topic but have so far at best made first attempts to explore the issues at a general level. Beyond the specialised literature on EU climate diplomacy and foreign climate policy (e.g., Adelle

DOI: 10.4324/9781003183037-1

et al. 2018), studies on general EU foreign policy have started to acknowledge the principal importance of the climate transition, in particular with respect to EU foreign energy policy and bilateral relations with selected partner countries such as Russia (Knodt et al. 2017; Godzimirski 2019). Research has also begun to explore the geopolitical implications of the global climate transition for the international distribution of power and interests and international interdependence (e.g., Scholten 2018; IRENA 2019; Van de Graaf and Sovacool 2020). And literature on environmental and climate security has highlighted the significance of both climate change and climate policy for security policy (Ide et al. 2021; Swatuk et al. 2021). However, these literatures have so far stayed at a rather general level, not exploring in much detail the wide-ranging implications of the European and global climate transition for EU external relations.

Against this background, this book aims to enhance our understanding of the concrete implications of decarbonisation for EU external relations and how these can be managed and reshaped. To this end, we analyse bilateral relations between the EU and its member states, on one side, and a selection of 'hard cases', namely fossil fuel exporting countries, on the other, with a twofold focus. First, we aim to identify in more detail the significant challenges that arise from the decarbonisation process in the specific context of the partner countries. Second, we endeavour to delineate the room and the opportunities for developing bilateral relations beyond fossil fuels under the particularly challenging conditions of the selected partner countries in order to (1) help advance the climate transition, and (2) foster generally fruitful bilateral relations. We therefore define the task of foreign relations under decarbonisation as twofold, namely as (1) facilitating the phase-out of fossil fuels and GHG emissions and (2) transitioning towards renewed sustainable external relations that are based on other sectors, including low-carbon ones, while preventing potentially destabilising effects of the transition.

This introductory chapter sets the scene for the overall volume in four steps. The next section first provides some further background on EU climate policy and external relations in the context of the global decarbonisation challenge. Subsequently, we discuss how our effort relates to and advances four related and evolving literatures on EU climate policy, EU foreign policy and external relations, environment/climate and security, and the geopolitics of energy. This is followed by an introduction of our analytical approach and methodology, including the criteria for the selection of the six case studies that lie at the core of this study. The final section provides a short overview of the remainder of the volume.

Background: decarbonisation, EU climate policy, and EU external relations

The decarbonisation challenge

In 2015, the international community adopted the Paris Agreement and set the course for global decarbonisation before the end of the century. On the basis of its aforementioned temperature goal of 2/1.5°C, the Paris Agreement establishes in its Article 4.1 the collective target of achieving net-zero GHG emissions in the second half of this century. Consequently, many countries have moved to cut and phase out GHG emissions. Crucial to these efforts is the transition away from fossil fuels to a decarbonised energy supply, but decarbonisation is also a much broader process that will require, for example, changes in land use, greater resource efficiency, and the preservation of carbon sinks, such as forests and oceans (IPCC 2014, 2018). Given the long-term nature of the climate transition, durable commitment and long-term planning for realising decarbonisation are required. Consequently, the Paris Agreement also foresees that its parties develop long-term strategies to this effect (Art. 4.19). These strategies complement medium-term national climate action plans called 'nationally determined contributions' (NDCs) towards the collective temperature and decarbonisation goals (on the Paris Agreement, see Doelle 2016; Falkner 2016; Klein et al. 2017).

Realising the global decarbonisation goal implies a twofold fundamental change in the way economies and societies work. First, it entails the decline and destabilisation of carbon-intensive development models based on the production and burning of fossil fuels and overexploitation of natural resources. Second, it requires the emergence of more sustainable forms of energy production and resource use. As a 'disruptive, contested, and non-linear' process (Geels et al. 2017), we may expect it to have both adverse and positive effects on countries, with the particular challenges and opportunities they face varying considerably between them. The scale of these challenges is likely to be linked to existing economic dependence on fossil fuels and other high-carbon assets, while the potential to reap benefits and take advantage of new opportunities arising from decarbonisation will depend on the flexibility and capacity of the economy and society to shift and diversify.

The climate transition constitutes a crosscutting challenge for countries' economies and societies. It cuts across many key sectoral systems, including transport, buildings, power, industry, agriculture, forestry, finance, and so on – constituted of ensembles of actors, technologies, infrastructures, economic structures, institutions, and ideas that produce resistance to change (Geels and Schot 2010). As a result, the climate transition requires countries to implement and manage deep, structural changes to their economies and societies transforming these key sectoral systems and overcoming their inherent resistance to change to effectively eliminate GHG emissions. The strength of governance and institutions, as well as overall economic and political stability, are therefore likely to be important success factors in driving deep decarbonisation processes. Conversely, conflict and instability can significantly undermine efforts to decarbonise (IPCC 2014, 2018).

EU climate policy towards decarbonisation

EU climate policy has developed significantly especially since the 2000s and has moved towards the aim of climate neutrality by 2050. Its roots go back at least to the early 1990s when the first legislative and policy action was taken in the context of the 1992 United Nations Framework Convention on Climate Change (UNFCCC). EU climate policy development accelerated in the 2000s, when the EU began to implement its GHG emission reduction commitment under the 1997 Kyoto Protocol (see Table 1.1) and established the EU Emissions Trading System (ETS). Since then, the development of EU climate policy has advanced further and has led to a rich acquis of climate legislation addressing all relevant sectors (including buildings, transport, industry, power, land use, etc.) and using a diversified array of regulatory, market-based, and procedural policy instruments. In this process, the EU has also stepwise strengthened its decadal GHG emission reduction objectives, as depicted in Table 1.1 (on the development of EU climate policy, see Jordan et al. 2010; Oberthür and Pallemaerts 2010; Skjærseth et al. 2016; Delbeke and Vis 2019; Kulovesi and Oberthür 2020).

In the late 2010s, the EU has firmly moved towards aiming for climate neutrality and, hence, full decarbonisation by 2050. Since 2019, the development of EU climate policy has proceeded under the umbrella of the European Commission's European Green Deal (European Commission 2019). This has included the strengthening of GHG emission reduction targets for 2030 and 2050 (see Table 1.1). Based on the Commission's strategic long-term vision and pathway to a climate-neutral economy (European Commission 2018), the European Council approved the 2050 climate neutrality target in December 2019, which formed the EU's long-term strategy submitted to the UNFCCC in March 2020 (European Union 2020). Both the 2050 climate neutrality target and the upgraded 2030 GHG emission reduction target of 55 percent are enshrined in the European Climate Law agreed by the Council of the EU and the European Parliament in April 2021. The evidence available so far suggests that the Covid-19 pandemic and especially the Next Generation EU Fund agreed in response may enhance the implementation of the European Green Deal and the climate targets (Dupont et al. 2020). Eventually, the climate neutrality target implies that EU consumption and imports of fossil fuels are set to decline steeply to reach residual levels by 2050.

The EU has also established overall headline targets for the share of renewable energy in final energy consumption and for the improvement of energy efficiency for 2020 (both 20 percent) and for 2030 (32 percent for renewable energy and 32.5 percent for energy efficiency). These targets for 2030 are expected to be further strengthened in the ongoing implementation of the upgraded GHG emission reduction target of at least 55 percent for 2030. The European Commission is scheduled to table related legislative proposals to this effect in mid-2021. Both energy efficiency improvements and the share of renewable energy will also have to be enhanced further beyond 2030 in order to achieve net zero emissions in 2050 (EEA 2020).

As mentioned above, the European Green Deal also entails an upgrading of 'climate policy integration', that is, the integration of climate policy objectives into other sectoral policies ranging from trade over industrial to agricultural and foreign policies (see Adelle and Russel 2013; Dupont 2016). The Green Deal aims at a new quality of climate policy integration, namely that no other EU policies do significant harm to the climate and sustainability transition pursued, but that all other EU policies support, and synergise with, this transition. In this context, it also acknowledges the social dimension of the climate transition through the launching of a Just Transition Mechanism designed to assist regions and sectors particularly dependent on fossil fuels and carbon-intensive industry – and it highlights EU external relations as an integral part of the European Green Deal, as discussed next (European Commission 2019).

The decarbonisation challenge for EU external relations

EU climate policy has a strong external and international dimension. The EU possesses a strong track record of international leadership on climate change, including a longstanding climate diplomacy strategy. Ever since the negotiations of the 1992 UNFCCC, the EU has been the most ambitious major economy in international climate politics and has

Table 1.1 EU greenhouse gas emission reduction targets

Year adopted (related UN climate development)	*GHG emission reduction target (base year 1990)*	*Status*
1990 (1992 UNFCCC)	Stabilisation by 2000 (CO_2 only)	Achieved
1997/1998 (1997 Kyoto Protocol)	8 percent by 2008–2012	Achieved
2007 (2009 Copenhagen COP 15)	20 percent by 2020	24 percent in 2019
2009 (2009 Copenhagen COP 15)	80–95 percent by 2050	Additional measures required
2014 (2015 Paris Agreement)	40 percent by 2030	Additional measures required
2019 (2021 Glasgow COP 26)	Climate neutrality by 2050	Additional measures required
2020 (2021 Glasgow COP 26)	55 percent by 2030	Additional measures required

Source: Oberthür and Dupont (2021: 1102).

aspired to act as an international leader. Over the years, it has developed and adapted its international strategy to become a 'leadiator' (leader cum mediator; see Bäckstrand and Elgström 2013). As such, it has been an important driving force behind the 1997 Kyoto Protocol and the 2015 Paris Agreement, including its further development post-2015. In general, the EU has advanced its domestic climate policy objectives in tandem with international climate policy developments (Oberthür and Dupont 2021; see also Table 1.1). Part of the development of the EU international leadership strategy has been an enhanced involvement of foreign services. For example, the EU established a Green Diplomacy Network in the early 2000s to coordinate national and EU foreign services in support of EU international climate policy. Since 2010, this has also involved the European External Action Service (EEAS) established under the Lisbon Treaty. In 2008, the Commission published a climate diplomacy strategy, to achieve a global climate agreement, support the implementation of climate policies in partner countries, and address the security risks arising from climate change in different regions around the world (European Commission 2008). Since 2015, the foreign/general affairs Council has furthermore elaborated and regularly updated a climate diplomacy action plan. And in 2019, a climate ambassador was appointed to the EEAS (Schunz 2019; Oberthür and Dupont 2021).

The European Green Deal accordingly contains a strong external dimension. Specifically, it aims at further profiling the EU as a global leader on climate change. To this end, it highlights a number of pertinent priorities, including further work under the Paris Agreement, continued and strengthened engagement with international partners bilaterally and in regional contexts (with emphasis on Africa and the EU neighbourhood), a further greening of trade policy, and enhancing synergy with the international financial system. In this context, it also explicitly recognises the role of climate change and environmental challenges as a significant threat multiplier and a source of instability (European Commission 2019: 20–22).

It remains to be seen whether the external dimension of the Green Deal will in reality go beyond a focus on how EU foreign policy can enhance 'climate diplomacy' to help advance the EU's international climate objectives, which has been prevalent so far (e.g., Federal Foreign Office 2019; for the United States, see Podesta and Stern 2020). The broader implications of climate change and climate policy for external relations and foreign policy at large have been slow to move into political focus. Hence, the 2019 mission letter to the incoming High Representative of the Union for Foreign Affairs and Security Policy, Josep Borrell, only mentioned climate change in passing (von der Leyen 2019). Only more recently, an acknowledgement of the geopolitical dimension of climate change and climate policy has come about, including with respect to planned policy measures such as a Carbon Border Adjustment Mechanism to level the playing field for EU industry vis-à-vis international competition arising from lower emission standards and the transition towards low-carbon energy sources (including hydrogen, batteries, and more; Timmermans and Borrel 2021).

However, the climate transition's implications for foreign policy have yet to receive fuller consideration in the development of EU external relations. Many questions remain un- (or under-) explored. What are the repercussions of the commitment to deep decarbonisation by the EU (and the world community) for the general relations with various partner countries? How could and should these relations be further developed and reshaped under decarbonisation? How can synergies with the decarbonisation agenda be fostered and trade-offs be minimised or managed? This book aims to make a contribution to bringing these questions into focus.

EU climate policy and external relations: four literatures

Aiming to advance our understanding of the repercussions of the European and global climate transition for EU external relations, this study in particular relates to and advances four literatures on the following: (1) EU climate diplomacy and foreign climate policy, (2) EU foreign policy and external energy relations, (3) environment/climate and security, and (4) the geopolitics of energy under decarbonisation. Here, we briefly discuss these literatures and how this study relates and contributes to them. Like the broader political debate addressed in the previous section, the academic debates about foreign policy and international relations in general and EU foreign policy and external relations in particular have only just begun to explore the implications of deep decarbonisation that are hence in much need of further investigation.

First, literature on EU climate diplomacy and foreign climate policy and governance has advanced over more than two decades. It has been embedded in the broader literature on general EU climate policy and governance that has explored a variety of aspects such as policy innovation dynamics, particular policy instruments such as the EU ETS, policy durability/robustness, the overall ambition and effectiveness, as well as the climate policy integration into other policy fields (e.g., Skjærseth et al. 2016; Delbeke and Vis 2019; Jordan and Moore 2020; Oberthür et al. 2021). The more specific literature on EU external climate policy has included analyses of the EU's longstanding and evolving international climate leadership; its multilateral, regional, and bilateral climate diplomacy; and the external effects of its domestic climate policies (e.g., Wurzel et al. 2017; Adelle et al. 2018; Oberthür and Dupont 2021; Petri and Biedenkopf 2021). This literature has remained largely focused on the role and effectiveness of the EU in affecting *climate* policy developments beyond its borders, be it at the multilateral level or transnational, regional, national, or subnational levels. This study complements and broadens this focus. We go beyond a focus on climate policy developments and explore the effects of the EU's climate transition, in the context of the global drive towards decarbonisation, on the general bilateral relations with other countries, especially fossil fuel exporters, as well as the scope for fruitfully reshaping these bilateral relations under these circumstances. While staying focused on the climate transition, we also expand our existing focus on *climate* diplomacy (and how it could support the transition) towards general diplomacy and EU bilateral/external relations (and how these may be able to cope with the effects of the transition).

A second point of reference for our investigation is the literature on EU foreign policy and external relations in general and on EU external energy policy in particular. This literature has investigated the shapes, procedures, and conditions of EU foreign policy as a field characterised by strong EU member-state competences (e.g., Keukeleire and Delreux 2014; Bickerton 2015; Jørgensen 2015). Discussions on EU external energy policy have focused on issues of the security of energy supply – with a major emphasis on Russia as a main supplier of not only gas but also oil and coal; dynamics of competition, cooperation, and coordination between the European Commission (as the main supranational actor) and EU member states; and the form and nature of EU external energy policy at the interface of market imperatives and a stronger strategic orientation (e.g., Goldthau and Sitter 2015; Herranz-Surrallés et al. 2020; Thaler 2020; Siddi and Kustova 2021). Considerable attention has been paid in this context to prominent and possibly paradigmatic cases such as the construction of the North Stream 2 pipeline connecting Russia and Germany and, more generally, pipeline politics (including with respect to Ukraine and Azerbaijan; ibid.). The climate agenda has begun to enter relevant analyses as an issue that shapes actors' interests and options, with few contributions exploring external cooperation on related aspects such as energy efficiency and renewable energy (e.g., Boute 2013; see also Casier 2015; Knodt et al. 2017; Godzimirski 2019). Overall, however, it remains fair to say that climate change and climate policy have not been at the core of this stream of literature, taken into account mainly as an intervening factor affecting to some extent the actual issue of interest. In contrast, this study puts climate change and climate policy centre stage in the investigation of EU external (energy) relations.

Third, the scholarship on environment/climate and security is relevant for our endeavour. While not focused on the EU and EU policy, this literature has highlighted, and provided evidence, that environmental problems and climate change are important threat multipliers (Gemenne et al. 2014; Ide et al. 2021; Swain et al. 2021). Hence, it has found the impacts of climate change – such as drought, water scarcity, or climate-induced migration – to heighten the risk of violent conflict emerging both within and across borders. In so doing, it has drawn attention to how heightened environmental stress and impacts interact with other factors. For example, climate change is considered to exacerbate existing tensions or lead to the emergence of new security threats especially in fragile, conflict-prone regions (ibid.; Rüttinger et al. 2015). More recently, this literature has expanded its scope beyond environmental impacts to also consider the effects of climate *policy* on the emergence of conflict. This has prominently included a focus on 'maladaptation' to the impacts of climate change resulting, for example, in the displacement of communities from traditional lands as a result of climate policies (e.g., Gemenne et al. 2014; Swatuk et al. 2021). Our investigation shares with this literature the interest in (1) exploring the effects of climate policy beyond its confines, namely on broader external relations, and (2) paying attention to other, contextual factors (including state fragility and the security context), with which climate policy-induced changes will interact to shape both future conditions in partner countries and the latters' relations with the EU (see below).

While not focusing on security and conflict, this study therefore also provides a rich set of case studies in which the interaction between climate policy and other key factors central in the investigation of climate and security is explored.

Fourth, our analysis relates and contributes to the evolving literature on the geopolitics of energy under conditions of the advancing climate transition. This literature has generally begun to explore the effects of changes resulting from the global move towards decarbonisation on the international distribution of power and interdependence against the backdrop of geographical factors and resource endowments (Steinmetz 2012; Ivleva and Tänzler 2019). Important aspects that have received particular and increasing attention in related sub-literatures concern the impact of growth of renewable energy, the resulting decline of fossil fuels, and the emerging rise of hydrogen as an energy carrier (e.g., de Jong et al. 2017; Scholten 2018; IRENA 2019; Van de Graaf and Sovacool 2020; Van de Graaf et al. 2020; Blondeel et al. 2021). The focus of this strongly emerging literature has been on broad and general effects across different groups and types of countries within their geographical contexts (including the availability of natural resources), as raw materials and potentials for renewables and hydrogen replace fossil fuel reserves and infrastructure as the key 'currency' of energy geopolitics. The literature has so far operated at a fairly aggregate level, with few country case studies on the more concrete implications. Our study adds such a concrete country-level analysis of the likely implications of the climate transition (taking into account geographical context and resource endowments) so as to identify the challenges and opportunities arising for EU relations with the fossil fuel exporting countries concerned. Such a focus on bilateral relations also nuances the focus on interdependencies prevailing in the geopolitics perspective. As such, our study may serve as an input into efforts to concretise, nuance, and scale up geopolitical implications (in terms of power and interdependence; see also Chapter 8).

Overall, our study builds on and contributes to the four aforementioned literatures. Its added value in particular arises from its rooting in concrete country case studies, its broad focus on general bilateral relations between the EU and the partner countries (beyond climate and energy proper) and on exploring the effects of decarbonisation on these relations, and the embedding of climate and energy policy and relations in their respective, variable context. As such, our analysis operates at the intersection of, and links, scholarship on (EU) climate governance and (EU) foreign policy/external relations, while also drawing on and contributing to research on the environment/climate and security, and the geopolitics of energy. In the next section, we further specify our analytical approach and methodology to this end.

Analytical approach and methodology

Our analytical approach and methodology follow from the overall purpose of our investigation. As indicated above, we especially aim at understanding the main challenges and opportunities that arise from the European and global climate transition for the bilateral relations of the EU and its member states. This implies seeking answers to several key questions. How is decarbonisation going to affect the relationship between the EU and the partner country? What are the likely repercussions on the partner country given the existing adaptive capacity and the broader context? What toeholds and potentials exist in areas other than fossil fuels and high-carbon products for advancing the climate transition and fruitfully developing bilateral relations under decarbonisation?

These questions also indicate the fundamentally exploratory nature of our study. Given the current state of research in the field, the study does not aim at explanation or hypothesis testing. Instead, we explore the aforementioned questions with respect to a cross-section of fossil fuel exporting countries to generate first insights which may subsequently be further detailed and investigated in additional studies. In the following, we discuss in more detail the rationale of our case selection as well as key features of our approach, which shape the structure of the case studies.

Case selection: a cross-section of 'hard' cases

To explore our overarching questions, we base our study on six carefully selected country case studies. As discussed above, such concrete case studies have so far been lacking to explore in further detail the impacts of decarbonisation on external relations. A case study approach allows to trace the varying circumstances in the countries under investigation as well as the differing starting points and potentials of EU bilateral relations with them. It also provides a basis for exploring the variation of conditions, contexts, and prospects across different countries.

In the case study selection, we first decided to focus on fossil fuel exporting countries. These countries are likely to face particular economic and political challenges arising from decarbonisation in the coming years and decades. Given their dependence on exports of high-carbon assets, and thus on other countries' demand for such commodities, their economies face particular strain, government budgets are set to experience shortfalls, and planning processes become more uncertain (Van de Graaf and Verbruggen 2015). Fossil fuel exporting countries also face the prospect of their substantial fossil fuel-related assets, including untapped hydrocarbon reserves and relevant infrastructure, becoming stranded by being prematurely devalued, written down or converted into liabilities (Carbon Tracker Initiative 2015; Ivleva et al. 2017; Curtin et al. 2019). As a result, EU external relations with these countries are also likely to face particular challenges, especially where a significant share of fossil fuel exports has so far gone to the EU. We thus focus on 'hard' cases where challenges for EU external relations can be expected to be significant and opportunities, in turn, to be rather limited.

We made the selection from a bigger sample of about 25 fossil fuel exporting countries derived from large country-level data sets showing economic dependence on fossil fuels and other high-carbon assets. Key indicators used were fossil fuel rents as a percentage of gross domestic product (GDP) and fossil fuel exports as a percentage of merchandise exports (World Bank 2021a, 2021b, 2021c and 2021d).

Second, within the group of hard cases, we decided to focus on 'second-tier' countries. First of all, a substantial body of literature and analysis has already addressed the major fossil fuel exporters to the EU, and in particular Russia – even if a stronger focus on the implications of decarbonisation has remained lacking (e.g., Youngs 2014; Goldthau and Sitter 2015; Khrushcheva and Maltby 2016; Rossbach 2018). Furthermore, putting emphasis on second-tier countries provides the opportunity to investigate, and substantiate, whether European and global decarbonisation constitutes a key challenge for EU external relations beyond the most prominent fossil fuel exporters within the EU neighbourhood and beyond, including – next to Russia – Norway and Algeria but also Saudi Arabia (see also Casier 2015).

Given the exploratory nature of our investigation, we furthermore attempted to ensure that the country selection would capture a variety of country circumstances that may be expected to affect the challenges and opportunities for EU external relations. We therefore tried to ensure a wide geographical spread, to capture different regional dynamics, and took into account the countries' trading relationship with the EU, especially regarding fossil fuels and other high-carbon assets. We also tried to ensure that fossil fuel exporters selected varied widely with respect to fragility and existing security threats, human development, strength of governance, climate change impacts, and development of sustainable energy sources. To this end, we considered a number of relevant indexes, including the Fragile States Index (Fund for Peace 2021), the Human Development Index (UNDP 2020), the Worldwide Governance Indicators (World Bank 2020a), the index of Climate Change Vulnerability (ND-GAIN 2020), and the Regulatory Indicators for Sustainable Energy (World Bank 2020b).

On the basis of these factors and indicators, we made a selection of six fossil fuel exporting countries with varying vulnerability to deep decarbonisation: Azerbaijan, Canada, Colombia, Indonesia, Nigeria, and Qatar. Of these, Nigeria and Indonesia can be considered highly vulnerable to the climate transition. Nigeria's economy is highly dependent on gas and particularly oil, since these two commodities account for almost all its export revenues (with the EU as an important export market). The country also scores poorly on indexes assessing state fragility, human development, strength of governance, and preparedness for climate change impacts, indicating that it may be ill-equipped to respond to economic shocks resulting from decarbonisation. Indonesia is a rapidly developing economy, experiencing huge growth in the demand for goods and energy. This growth is highly dependent on the exploitation of its abundant hydrocarbons, primarily coal and oil, as well as other high-carbon assets, such as palm oil. While trade in fossil fuels between Indonesia and the EU has been negligible, the EU has become a major importer of Indonesian palm oil. Indexes on state fragility, human development, strength of governance, and preparedness for climate change impacts, all suggest a medium to high level of concern.

At the other end of the spectrum, Canada and Qatar can be deemed comparatively less vulnerable and well placed to actively adapt to and shape decarbonisation. As reflected in the aforementioned indexes, Canada is a stable, highly developed democracy with a diversified economy and a well-educated workforce. Although its oil industry remains economically important, Canada has an established climate policy architecture and renewable energy sources are at an advanced stage of development. Trade in fossil fuels only accounts for around 10 percent of trade with the EU, and it has well-established relations with the EU in many areas, including in domains that support decarbonisation. Qatar is a rich petrostate with stable governance structures and one of the highest per capita incomes in the world. Although it remains highly dependent on fossil fuels, it has used the wealth generated by its oil industry to develop its gas sector, which has longer-term prospects under decarbonisation, as well as other areas of its economy and its external relations.

The remaining two countries, Colombia and Azerbaijan, face both major risks and opportunities as the world decarbonises. In Colombia, revenues from coal and oil exports are economically significant, but the economy is increasingly diversifying into other areas and it has a strong climate policy architecture to steer growth into low-emission areas. However, the country is still emerging from decades of internal conflict and lasting reconciliation is likely to rely on political and economic stability over the coming decades. With the EU being a key market for Colombian coal at present, future bilateral relations may likely rely on developing trade in other sectors. Azerbaijan's economy is also highly dependent on oil and gas, and among the sample countries chosen it is the country that is most dependent on trade in these commodities with the EU. Scores on human development, governance, and climate change impacts indicate low or modest levels of concern, although resilience to external shocks may be limited.

Overall, our selection represents a broad cross-section of fossil fuel exporting countries likely to be vulnerable to varying extents to falling demand for coal, oil, gas, and other high-carbon products resulting from the global climate transition. This selection should serve well the purpose of our study to explore the varying challenges and opportunities arising for EU external relations from decarbonisation. Given the study's explorative nature, the case selection did not aim at systematic variation on any 'dependent' or 'independent' variables but aimed at variation in both sample countries' vulnerability and their existing bilateral relations with the EU, with both to be further specified in the case studies. Focusing on a key risk group under decarbonisation, the country selection reflects a broad cross-section in terms of state stability or fragility, economic development, geographical region, and the relative reliance on coal, oil, and gas.

Paying attention to context and wider external relations

To some extent already reflected in the criteria of case selection discussed above, our analytical approach firmly rests on paying attention to relevant contextual conditions and broader external relations beyond fossil fuels and energy. On the one hand, the challenges and opportunities arising from decarbonisation for countries in general and fossil fuel exporters in particular depend on wider economic and political framework conditions, both internal and external. On the other hand, EU foreign policy can both help address the challenges and foster the opportunities so as to advance decarbonisation and advance bilateral relations under decarbonisation. Such potential of foreign policy exists beyond the area of climate and energy. We here detail our approach to capturing these two aspects further.

As regards broader economic and political framework conditions, we suggest focusing on key aspects beyond sectors directly affected by reductions in demand for fossil fuels and other high-carbon goods. Climate action and decarbonisation processes are likely to have broader effects, including on trade relations, finance and investment flows, official development assistance, or efforts to promote conflict prevention and peacebuilding. Such a broader perspective entails consideration of various other fragility and security risks that may intersect with the impacts of decarbonisation, including existing political tensions or security risks at national or regional levels (e.g., Kim and Conceição 2010), strength of governance (that may strongly affect the ability of governments to diversify away from fossil fuels: Esanov 2012; Collier 2012), and climate change impacts likely to act as a 'threat multiplier' in the coming decades (Rüttinger et al. 2015) and potentially affecting important economic sectors (such as agricultural production) as well as food, water, and livelihood security in general. By appraising the development and stability of economy, society, and political governance, we aim to investigate how the carbon-dependent economy may intersect with other potentially destabilising factors.

Furthermore, we consider the barriers to and potential for diversifying the economy. This encompasses an analysis of the country's climate politics and policy, including its international engagement under the UNFCCC and the Paris Agreement, the status of domestic climate policy, the political influence of fossil fuel industries, and the level of public support for climate action. Beyond climate politics, the potential for successfully managing the climate transition will also depend on factors such as the current levels of development of low-carbon sectors (e.g., renewable energies), economic diversification, and strength of the educational system. Overall, we thereby capture key factors affecting partner countries' exposure to global decarbonisation and their ability to successfully manage the climate transition. As such, understanding these challenges and potentials is also key for identifying the scope for fruitfully developing bilateral relations.

Furthermore, gauging the prospects of EU external relations with the indicated countries under decarbonisation requires understanding the existing bilateral relations. One may be tempted to focus on bilateral relations in climate and energy policy, because these are most closely related to the climate transition. However, we consider that a broad approach to foreign policy seems most appropriate for our purpose of assessing the challenges and, importantly, the opportunities for developing EU external relations with the selected partner countries in general under decarbonisation. Consequently, we attempt to capture the main tenets of EU relations with and policy towards the partner countries concerned, with a particular focus on underexploited potentials. While not necessarily capturing external relations comprehensively, the following five policy fields stand in focus in the country case studies as key areas of general foreign policy that are particularly implicated by deep decarbonisation:

- *Climate and energy policy* lies at the centre of the climate transition and is hence directly implied. As trade in fossil fuels fades under decarbonisation, the promotion of climate friendly solutions (in the field of energy production and beyond) and key climate policy instruments (such as the EU ETS) constitute possible alternatives for EU external relations.
- *Trade and investment* are at the centre of the EU's external relations and 'market power' (Damro 2014). Related agreements may be further recalibrated to strengthen support for global decarbonisation and the climate transition of partner countries in the context of the European Green Deal. As a key area of EU external relations, trade and investment can serve to develop alternatives to trade in fossil fuels and other high-carbon products.
- *Research and education* constitute another prominent area of EU external relations with partner countries. They are key to much-needed efforts at developing a more knowledge-based economy and diversifying economies away from fossil fuels and high-carbon products. Instruments at the disposal of the EU may range from

cooperation programmes for higher education to the provision of financial and technical assistance for vocational education and training to supporting the development of skills required in low-carbon sectors.
- *Security policy* is perhaps the highest politics core of any external relations. In an era of decarbonisation, it can specifically help address destabilisation risks related to the climate transition, which may shape evolving regional power relations and coalitions. We therefore consider the potential of EU security policy to contribute to long-term engagement strategies and addressing partner countries' risks.
- *Development finance and cooperation* is another key area of EU external relations with eligible recipient countries. Taken together, the EU and its member states are by far the biggest source of development assistance worldwide, providing €75.2 billion to developing countries in 2019 (European Commission 2020). As this is increasingly used to support climate action, development cooperation can help accelerate the just transition towards a decarbonised economy in partner countries, thereby nurturing alternatives to fossil fuels.

While these policy areas delimit the overall scope of our investigation of the status and potential of EU external relations with its partner countries in focus, this investigation takes into account the particularities of each partner country. Hence, development cooperation is only explored for countries eligible for related support. Similarly, cooperation on research and education and climate and energy may have varying potential across the selected countries and are thus explored to varying extents. At the same time, the assessment of EU external relations with partner countries concerned generally includes also a more general, cross-cutting dimension of political relations.

A clarification on the scope of EU foreign policy under investigation in this study is in order. EU member states remain key actors in external relations. The competences and roles of the EU itself and its member states fluctuate per policy field. The supranational EU role is strongest in areas of exclusive EU competence, in particular in trade. In contrast, member states remain predominant especially with respect to security policy. The other areas are characterised by varying mixes of competence, so that both the EU itself and its member states play important roles. While a full assessment of EU member states' foreign policy towards the countries selected is beyond the scope of this study and we put emphasis on the EU level, we try to reflect the reality of both the EU and the member states being engaged in, and together shaping, European foreign policy in an aggregate way. For ease of reference, we generally refer to 'the EU' throughout, even though the analysis is aimed to capture, and relate to, 'the EU and its member states'. We occasionally refer to 'the EU and its member states' where the role of member states is particularly in focus. We return to the scope for further deepening the analysis with respect to EU member states in the concluding chapter.

Structure and methods of the case studies

Based on these considerations, we devised a common structure of the country case studies in order to identify the barriers and risks the countries are facing under decarbonisation and, more generally, to analyse the countries' potential for moving beyond fossil fuels and to scrutinise existing bilateral relations with the EU and their promise. Accordingly, each case study consists of the following five steps.

First, we investigate the partner country's exposure to decarbonisation in the context of wider fragility and security risks. Specifically, case studies examine how dependent a country's economy and trade are on fossil fuels or other high-carbon goods and how exposed the economy of a country could be to reductions in demand for these goods, both globally and within the EU. As discussed above, we also broaden the scope of our analysis beyond sectors directly affected by such demand reductions. We assess the development and stability of economy, society and political governance, including internal and regional security risks as well as the potential (security) risks arising from climate change itself.

Second, and in line with the discussion above, the country case studies explore partner countries' past and present efforts to decarbonise. This analysis includes partner countries' emission profiles, the status of domestic climate policy (including the strength of institutions in charge of implementing it), and the role in the UNFCCC and the Paris Agreement, most importantly including its NDC. It also addresses key components of domestic climate politics, in particular the political weight and influence of the fossil fuel industries as well as the level of public support for climate action. This analysis lays a basis for assessing the scope for deepening cooperation on climate and energy.

Third, the case studies investigate existing trends and potentials in key areas such as economy and education in order to gauge countries' potential to move beyond fossil fuels and high-carbon products and the scope for expanding related cooperation. We in particular consider potential in key low-carbon sectors such as renewable energy and energy efficiency, including related policy support. Moreover, education and skills development and efforts to diversify the economy receive specific attention. We remain attentive to specific country characteristics in the sectoral focus and emphasis of our analysis (e.g., analysing forestry and agriculture where appropriate).

Fourth, our case studies examine the current status of countries' external relations with the EU. Starting from the general institutional framework of cooperation and exchange, we focus on the five key areas highlighted above (climate and energy policy, trade and investment, research and education, security policy, development finance and cooperation). This analysis aims to unearth the existing seeds of a relationship beyond fossil fuels and high-carbon products. It puts varying emphases depending on the relevance of individual key areas for EU relations with the specific country in question, for example, regional security, development cooperation, or the broader cooperation agenda. This analysis can inform thinking about how relations can fruitfully be developed under decarbonisation.

The case studies conclude with a synthesis of the main findings. These also serve as a basis for identifying focal points for the fruitful development of EU external relations with the country in question in the context of the climate transition. We hence try to advance towards ascertaining the scope for and promise of fruitful EU foreign policy under decarbonisation.

In line with the explorative approach of the study, the case studies chosen are generally based on in-depth desk research deploying a variety of sources. They are rooted in the extensive study of available secondary literature (both academic and 'grey') on the relevant aspects of the countries' characteristics, economics and politics, complemented by selective and focused document analysis and broad examination of news reports. In addition, our analysis is built on a range of indicators and databases measuring the level and trajectory of socioeconomic development, political stability and strength of governance, and vulnerability to climate change impacts, as well as assessments of national and regional security challenges.

The research on the case studies was generally completed in late 2019 and updated in 2020/2021. Our case studies do not therefore include a detailed analysis of the impacts of the Covid-19 pandemic, although we have attempted to include limited and cautious updates where feasible, bearing in mind that the exact implications of the pandemic continue to unfold (and will continue to unfold for some time to come). If anything, the pandemic-induced economic crisis and consequent decline of fossil fuel prices have further reinforced the findings related to the vulnerability of the case study countries to price fluctuations and the challenges, need, and urgency of moving beyond fossil fuels. We are confident that the analysis presented here remains generally relevant also in the post-pandemic world – even if the fallout of the pandemic may require some adaptations.

Overview of the study

The remainder of this book implements the aforementioned plan. The case studies on the six selected countries highly dependent on the export of fossil fuels or other high-carbon products form the core of the present study. The case studies are presented roughly in the order of the fragility and security risks that the countries face, taking into account a number of indicators (see above). As the country that is subject to the greatest fragility and security risks, the Nigeria case study is hence presented first, followed by Indonesia, Azerbaijan, Colombia, Qatar, and Canada. As mentioned above, the case studies follow a common structure and template so as to systematically tease out the challenges and opportunities that decarbonisation will present for these countries and, specifically, for EU relations with them. The common structure and template also facilitate the aggregation and comparison of the results in the final chapter. In the following, we present a brief preview of each chapter.

Chapter 2 addresses Nigeria as the country most fragile and vulnerable to decarbonisation in our selection. The fortunes of Nigeria's economy are closely tied to the oil price, making it vulnerable to oil market fluctuations and the phase-out of fossil fuels. While its oil and gas reserves have brought huge wealth to the country, the profits and benefits have not been equitably shared or sufficiently invested in key public services and infrastructure. The chapter demonstrates that this underinvestment coupled with other major challenges has held back efforts to diversify the economy, despite significant existing potential. Although efforts to increase and liberalise EU-Nigeria trade in non-oil sectors have stalled, there is plenty of scope for the EU to support Nigeria in sustainably developing its non-oil economy. There are numerous entry points for political dialogue and development cooperation to enable the EU to advance its relations with Nigeria under decarbonisation. These relate to Nigeria's many pressing security, economic, and socio-political issues and to the promotion of a decarbonised economy.

Chapter 3 turns to Indonesia and EU-Indonesia relations. Indonesia's economic growth has been underpinned by a huge variety of natural resources at its disposal. After the Amazon, it is the world's most heavily forested region, and it also has rich deposits of hydrocarbons and minerals. Many of these natural assets have also been developed for export. The country is the second largest exporter of coal after Australia and the world's largest exporter of palm oil – the latter accounting for a large proportion of its trade with the EU. Indonesia has diversified its export base beyond fossil fuels over the past 15 years, but it has also become increasingly dependent on coal to meet rising domestic energy demand. Overall, Indonesia is an increasingly important partner for the EU, as it is both a major economy and a leading voice within the Association of Southeast Asian Nations (ASEAN). The chapter argues that the EU has established a fruitful basis for cooperation, with ongoing negotiations on a comprehensive economic partnership agreement enhancing the potential for further cooperation, including in areas that are relevant for decarbonisation.

Chapter 4 analyses Azerbaijan and the prospect of EU relations with it under decarbonisation. Azerbaijan is highly dependent on the production and export of oil and gas. Accordingly, while the development of its oil and gas resources has in general helped to advance welfare and has provided for political stability in the twenty-first century, its economic development has fluctuated with the world market price of oil and gas. Efforts to diversify the economy have had limited effects so far but provide an important entry point for developing EU-Azerbaijan relations beyond fossil fuels. Azerbaijan also faces serious issues regarding state legitimacy and repression as well as security challenges in the context of its conflict with Armenia over Nagorno-Karabakh. The case study demonstrates that EU-Azerbaijan cooperation can build on a firm and broad institutional framework. The EU is Azerbaijan's most important trading partner, accounting for nearly half of the latter's trade. While energy and especially oil and gas have been key to EU-Azerbaijan relations, cooperation on both renewable energy and education holds significant potential.

Colombia and EU-Colombia relations are in focus in Chapter 5. Fossil fuel export revenues make a significant contribution to Colombia's economy, which has been vulnerable to drops in commodity prices. However, the past 25 years have also seen important structural transformations and shifts towards economic diversification. The 2016 peace agreement that ended the internal conflict with the Revolutionary Armed Forces of Colombia (FARC) considerably improved Colombia's political stability, although a stable economy and continued political commitment are needed to ensure lasting peace. Rural development, climate action, and economic diversification are important entry points for EU-Colombia cooperation to support both decarbonisation and post-conflict development processes. Overall, the chapter argues that the EU can build on a sound basis for cooperation, with increasing political dialogue in relevant areas of mutual interest. The EU was a valuable partner in the final stages of the peace process and possesses significant potential to tackle the root causes of the conflict as a donor as well as via investment and trade.

Qatar is the focus of Chapter 6. Qatar is highly dependent on the production and export of oil and gas, which form the backbone of its economy and the basis of its considerable wealth. Qatar's reliance on oil and gas also makes it vulnerable to price fluctuations and the phase-out of fossil fuels. Efforts to diversify its economy have been effective only to a limited extent so far. These efforts and other circumstances (e.g., the geopolitical position of Qatar in the Gulf region and the Middle East more generally) provide important entry points for developing EU-Qatar relations beyond fossil fuels. EU-Qatar relations have grown closer over the years, and a Cooperation Agreement was signed in 2018. Furthermore, EU-Qatar relations are embedded in broader cooperation between the EU and the Gulf Cooperation Council. The chapter argues that the important and evolving areas of trade and investment as well as energy possess significant potential to integrate decarbonisation (e.g., through developing Qatar's significant renewables potential). Given the precarious situation in the Middle East, security and geopolitics form another key area of the relations between Qatar and the EU.

Chapter 7 examines Canada and EU-Canada relations under decarbonisation. As the production and export of oil, gas, and coal are important for Canada's economy, the country is vulnerable to price fluctuations and the phase-out of fossil fuels. However, with a thriving and highly developed economy that goes far beyond fossil fuels, Canada has the potential to build on a highly diversified economy to transition towards climate neutrality. Regional differences provide a significant challenge for this process. The chapter argues that historical ties and shared interests and values provide a fruitful basis for developing EU-Canada relations under decarbonisation. These relations have been close and dense for several decades, as evidenced by the conclusion of the Comprehensive Economic and Trade Agreement (CETA) in 2016, as well as the long-established close cooperation in a multitude of settings and areas. Trade and investment as well as energy (both fossil fuels and renewables) are important and evolving areas, with significant potential to integrate decarbonisation. In addition, both the EU and Canada face similar challenges in the climate transition, including with respect to the just transition of regions and sectors highly dependent on the fossil fuel-based economy.

The concluding chapter, Chapter 8, first of all synthesises the main findings of all the case studies. The chapter specifically highlights that decarbonisation generally constitutes a key political and economic challenge for fossil fuel exporters that regularly intersects with other, specific fragility and security risks. This challenge is further aggravated by regularly comparatively underdeveloped climate policy frameworks and unfavourable climate politics. The development of relations with the EU can build on pre-existing arrangements of various strength and form, which generally provide ample potential and room for developing bilateral relations beyond fossil fuels. The chapter then explores the broader contributions of the study to advancing the relevant existing literatures discussed previously. It concludes with the identification of seven avenues for future research on EU external relations under decarbonisation.

Notes

1 There is an important difference in the EU context between 'foreign and security policy' and 'external relations'. However, for the purposes of this study, we use both 'foreign policy' and 'external relations' interchangeably.

References

Adelle, C. and Russel, D. (2013) 'Climate policy integration: A case of déjà vu?', *Environmental Policy and Governance*, 23(1), pp. 1–12.

Adelle, C., Biedenkopf, K., and Torney, D. (eds.) (2018) *European Union external environmental policy: Rules, regulation and governance beyond borders*. Houndmills: Palgrave Macmillan.

Bäckstrand, K. and Elgström, O. (2013) 'The EU's role in climate change negotiations: From leader to "lediator"', *Journal of European Public Policy*, 20(10), pp. 1369–1386.

Bickerton, C.J. (2015) *European Union foreign policy: From effectiveness to functionality*. London: Palgrave Macmillan.

Blondeel, M., Bradshaw, M.J., Bridge, G., and Kuzemko, C.(2021) 'The geopolitics of energy system transformation: A review', *Geography Compass*, e12580. https://doi.org/10.1111/gec3.12580.

Boute, A. (2013) 'Energy efficiency as a new paradigm of the European external energy policy: The case of the EU–Russian Energy Dialogue', *Europe-Asia Studies*, 65(6), pp. 1021–1054.

Carbon Tracker Initiative (2015) *The $2 Trillion Stranded Assets Danger Zone: How Fossil Fuel Firms Risk Destroying Investor Returns*. Available at: http://www.carbontracker.org/wp-content/uploads/2015/11/CAR3817_Synthesis_Report_24.11.15_WEB2.pdf (accessed 29 June 2021).

Casier, T. (2015) 'The geopolitics of the EU's decarbonization strategy: A bird's eye perspective'. In: Dupont, C. and Oberthür, S. (eds.) *Decarbonization in the European Union. Energy, climate and the environment*. London: Palgrave Macmillan, pp. 159–179.

Collier, P. (2012) 'The political economy of natural resources', *Social Research*, 77(4), pp. 1105–1132.

Curtin, J., McInerney, C., Gallachóir, B.O., Hickey, C., Deane, P., and Deeney, P. (2019) 'Quantifying stranding risk for fossil fuel assets and implications for renewable energy investment: A review of the literature', *Renewable and Sustainable Energy Reviews*, 116. https://doi.org/10.1016/j.rser.2019.109402.

Damro, C. (2014) 'Market power Europe', *Journal of European Public Policy*, 19(5), pp. 682–699.

de Jong, S., Auping, W.L., Oosterveld, W., Usanov, A., Abdalla, M., Van de Bovenkamp, A., and Frattina della Frattina, C. (2017) *The Geopolitical Impact of Climate Mitigation Policies: How Hydrocarbon Exporting Rentier States and Developing Nations Can Prepare for a More Sustainable Future*. The Hague Centre for Strategic Studies. Available at: https://hcss.nl/report/the-geopolitical-impact-of-climate-mitigation-policies/ (accessed 28 June 2021).

Delbeke, J. and Vis, P. (eds.) (2019) *Towards a climate-neutral Europe: Curbing the trend*. Abingdon/New York: Routledge.

Doelle, M. (2016) 'The Paris Agreement: Historic breakthrough or high stakes experiment?', *Climate Law*, 6(1/2), pp. 1–20.

Dupont, C. (2016) *Climate policy integration into EU energy policy: Progress and prospects*. Abingdon: Routledge.

Dupont, C., Oberthür, S., and von Homeyer, I. (2020) 'The Covid-19 crisis: A critical juncture for EU climate policy development?', *Journal of European Integration*, 42(8), pp. 1095–1110. https://doi.org/10.1080/07036337.2020.1853117.

EEA – European Environment Agency (2020) *Trends and Projections in Europe 2020: Tracking Progress towards Europe's Climate and Energy Targets. EEA Report No 13/2020*. Copenhagen: EEA. Available at: https://www.eea.europa.eu/publications/trends-and-projections-in-europe-2020 (accessed 29 June 2021).

Esanov, A. (2012) *Diversification in Resource-Dependent Countries. Its Dynamics and Policy Issues*. Revenue Watch Institute. Available at: https://resourcegovernance.org/sites/default/files/documents/rwi_econ_diversification_intro1.pdf (accessed 29 June 2021).

European Commission (2008) *Climate Change and International Security. Paper from the High Representative and the European Commission to the European Council. 14.03.2008. S113/08*. Available at: https://www.consilium.europa.eu/uedocs/cms_data/docs/pressdata/en/reports/99387.pdf (accessed 20 June 2021).

European Commission (2018) *A Clean Planet for All: A European Strategic Long-Term Vision for Prosperous, Modern, Competitive and Climate Neutral Economy. 28.11.2018. COM(2018) 773 Final*. Available at: https://eur-lex.europa.eu/legal-content/EN/TXT/PDF/?uri=CELEX:52018DC0773 (accessed 20 June 2021).

European Commission (2019) *The European Green Deal. 11.12.2019. COM(2019) 640 final*. Available at: https://ec.europa.eu/info/sites/default/files/european-green-deal-communication_en.pdf (accessed 20 June 2021).

European Commission (2020) *The European Union Remains World's Leading Donor of Official Development Assistance with € 75.2 Billion in 2019. 16.04.2020. Press Release*. Available at: https://ec.europa.eu/commission/presscorner/detail/en/ip_20_674 (accessed 29 June 2021).

European Union (2020) *Long-Term Low Greenhouse Gas Emission Development Strategy of the European Union and Its Member States. 5.03.2020. 6612/20*. Available at: https://data.consilium.europa.eu/doc/document/ST-6612-2020-INIT/en/pdf (accessed 29 June 2021).

Falkner, R. (2016) 'The Paris Agreement and the new logic of international climate politics', *International Affairs*, 92(5), pp. 1107–1125. https://doi.org/10.1111/1468-2346.12708.

Federal Foreign Office (2019) *Federal Foreign Office Climate Diplomacy Report*. Available at: https://www.auswaertiges-amt.de/blob/2296210/3371d7d8f8b282036ff0db19af1db021/200124-klimabericht-dl-data.pdf (accessed 29 June 2021).

Fund for Peace (2021) 'Fragile States Index'. Retrieved 20 June 2021 from https://fragilestatesindex.org.

Geels, F.W. and Schot, J. (2010) 'The dynamics of transitions: A socio-technical perspective'. In: Grin, J., Rotmans, J., and Schot, J. (eds.) *Transitions to sustainable development: New directions in the study of long term transformative change*. New York and London: Routledge, pp. 11–104.

Geels, F.W., Sovacool, B.K., Schwanen, T., and Sorrell, S. (2017) 'The socio-technical dynamics of low-carbon transitions', *Joule*, 1(3), pp. 463–479.

Gemenne, F., Barnett, J., Adger, W.N., and Dabelko, G.D. (2014) 'Climate and security: Evidence, emerging risks, and a new agenda', *Climatic Change*, 123(1), pp. 1–9.

Godzimirski, J.M. (ed.) (2019) *New political economy of energy in Europe. Power to project, power to adapt*. Cham: Palgrave Macmillan.

Goldthau, A. and Sitter, N. (2015) *A liberal actor in a realist world. The European Union regulatory state and the global political economy of energy*. Oxford: Oxford University Press.

Herranz-Surrallés, A., Solorio, I., and Fairbrass, J. (2020) 'Renegotiating authority in the Energy Union: A framework for analysis', *Journal of European Integration*, 42(1), pp. 1–17. https://doi.org/10.1080/07036337.2019.1708343.

Ide, T., Bruch, C., Carius, A., Conca, K., Dabelko, G.D., Matthew, R., and Weinthal, E. (2021) 'The past and future(s) of environmental peacebuilding', *International Affairs*, 97(1), pp. 1–16.

IPCC (2014) *Climate Change 2014: Synthesis Report. Contribution of Working Groups I, II and III to the Fifth Assessment Report of the Intergovernmental Panel on Climate Change*. Geneva: IPCC. Available at: https://www.ipcc.ch/site/assets/uploads/2018/02/SYR_AR5_FINAL_full.pdf (accessed 29 June 2021).

IPCC (2018) *Global Warming of 1.5°C. An IPCC Special Report on the Impacts of Global Warming of 1.5°C above Pre-Industrial Levels and Related Global Greenhouse Gas Emissions Pathways, in the Context of Strengthening the Global Response to the Threat of Climate Change, Sustainable Development, and Efforts to Eradicate Poverty*. Geneva: IPCC. Available at: https://www.ipcc.ch/site/assets/uploads/sites/2/2019/06/SR15_Full_Report_High_Res.pdf (accessed 29 June 2021).

IRENA – International Renewable Energy Agency (2019) *A New World: The Geopolitics of the Energy Transformation*. Available at: http://geopoliticsofrenewables.org/Report (accessed 29 June 2021).

Ivleva, D., Wolters, S., Scholl, C., Schlösser, T., and Schultze, K.R. (2017) *From Riches to Rags? Stranded Assets and the Governance Implications for the Fossil Fuel Sector. Discussion Paper.* Berlin: GIZ. Available at: https://www.adelphi.de/en/system/files/mediathek/bilder/From%20Riches%20to%20Rags%20-%20Stranded%20Assets%20and%20the%20Governance%20Implications%20for%20the%20Fossil%20Fuel%20Sector%20-%20GIZ%20adelphi.pdf (accessed 29 June 2021).

Ivleva, D. and Tänzler, D. (2019) 'Geopolitics of Decarbonisation: Towards an Analytical Framework'. *Climate Diplomacy Brief.* Berlin: adelphi. Available at: https://climate-diplomacy.org/sites/default/files/2020-10/Geopolitics-of-decarbonisation_Towards-an-analytical-framework.pdf (accessed 29 June 2021).

Jordan, A. and Moore, B. (2020) *Durable by design? Policy feedback in a changing climate.* Cambridge: Cambridge University Press.

Jordan, A., Huitema, D., van Asselt, H., Rayner, T., and Berkhout, F. (eds.) (2010) *Climate change policy in the European Union: Confronting the dilemmas of mitigation and adaptation?* Cambridge: Cambridge University Press.

Jørgensen, K.E. (2015) 'The study of European foreign policy: Trends and advances'. In: Jørgensen, K.E., Kalland Aarstad, A., Drieskens, E., Laatikainen, K., and Tonra, B. (eds.) *Handbook of European foreign policy.* Thousand Oaks, CA: Sage, pp. 14–29.

Keukeleire, S. and Delreux, T. (2014) *The foreign policy of the European Union.* 2nd ed. Basingstoke: Palgrave Macmillan.

Khrushcheva, O. and Maltby, T. (2016) 'The future of EU-Russian energy relations in the context of decarbonisation', *Geopolitics*, 21(4), pp. 799–830.

Kim, N. and Conceição, P. (2010) 'The economic crisis, violent conflict, and human development', *International Journal of Peace Studies*, 15(1), pp. 29–43. http://www.jstor.org/stable/41852999.

Klein, D., Carazo, M.P., Doelle, M., Bulmer, J., and Higham, A. (eds.) (2017) *The Paris Agreement on climate change: Analysis and commentary.* Oxford: Oxford University Press.

Knodt, M., Chaban, N., and Nielsen, L. (2017) *Bilateral energy relations between the EU and emerging powers: Mutual perceptions of the EU and Brazil, China, India and South Africa.* Baden-Baden: Nomos.

Kulovesi, K. and Oberthür, S. (2020) 'Assessing the EU's 2030 Climate and Energy Policy Framework: Incremental change toward radical transformation?' *Review of European, Comparative and International Environmental Law*, 29(2), pp. 151–66. https://doi.org/10.1111/reel.12358.

ND-GAIN (2020) 'Rankings'. Retrieved 29 June 2021 from https://gain.nd.edu/our-work/country-index/.

Oberthür, S., Jordan, A., and von Homeyer, I. (2021) 'EU climate and energy governance in times of crisis: Towards a new agenda', *Journal of European Public Policy*, 28(7), pp. 959–979. https://doi.org/10.1080/13501763.2021.1918221.

Oberthür, S. and Dupont, C. (2021) 'The European Union's international climate leadership: Towards a grand climate strategy?' *Journal of European Public Policy*, 28(7), pp. 1095–1114. https://doi.org/10.1080/13501763.2021.1918218.

Oberthür, S. and Pallemaerts, M. (eds.) (2010) *The new climate policies of the European Union: Internal legislation and climate diplomacy.* Brussels: VUB Press.

Petri, F. and Biedenkopf, K. (2021) 'Weathering growing polarization? The European Parliament and EU foreign climate policy ambitions', *Journal of European Public Policy*, 28(7), pp. 1057–1075. https://doi.org/10.1080/13501763.2021.1918216.

Podesta, J. and Stern, T. (2020) 'A foreign policy for the climate. How American leadership can avert catastrophe', *Foreign Affairs*, 05/06.2020. Retrieved 29 June 2021 from https://www.foreignaffairs.com/articles/united-states/2020-04-13/foreign-policy-climate.

Rossbach, N.H. (2018) *The Geopolitics of Russian Energy – Gas, Oil and the Energy Security of Tomorrow'. FOI-R--4623–SE.* Stockholm: Swedish Defence Research Agency. Available at: https://www.foi.se/rapportsammanfattning?reportNo=FOI-R--4623--SE (accessed 28 June 2021).

Rüttinger, L., Stang, G., Smith, D., Tänzler, D., Vivekananda, J. et al. (2015) *A New Climate for Peace – Taking Action on Climate and Fragility Risks*. Berlin/London/Washington/Paris: adelphi, International Alert, The Wilson Center, European Union Institute for Security Studies. Available at: https://climate-diplomacy.org/sites/default/files/2020-11/NewClimateForPeace_FullReport_small_0.pdf (accessed 12 July 2021).

Scholten, D. (ed.) (2018) *The geopolitics of renewables*. New York: Springer International Publishing.

Schunz, S. (2019) 'The European Union's environmental foreign policy: From planning to a strategy?' *International Politics*, 56(3), pp. 339–358.

Siddi, M. and Kustova, I. (2021) 'From a liberal to a strategic actor: The evolution of the EU's approach to international energy governance', *Journal of European Public Policy*, 28(7), pp. 1076–1094. https://doi.org/10.1080/13501763.2021.1918219.

Skjærseth, J.B., Eikeland, P.O., Gulbrandsen, L.H., and Jevnaker, T. (2016) *Linking EU climate and energy policies: Decision-making, implementation and reform*. Cheltenham: Edward Elgar Publishing.

Steinmetz, G. (2012) 'Geopolitics'. In: Ritzer, G. (ed.) *Encyclopedia of globalisation*. New York and London: Routledge, pp. 1–25.

Swain, A., Öjendal, J., and Jägerskog, A. (eds.) (2021) *Handbook of security and the environment*. Cheltenham: Edward Elgar Publishing.

Swatuk, L.A., Thomas, B.K., Wirkus, L., Krampe, F., and da Silva, L.P.B. (2021) 'The "Boomerang Effect": Insights for Improved Climate Action', *Climate and Development*, 13(1), pp. 61–67.

Thaler, P. (2020) *Shaping EU foreign policy towards Russia: Improving coherence in external relations*. Cheltenham: Edward Elgar Publishing.

Timmermans, F. and Borrell, J. (2021) 'The geopolitics of climate change', *Project Syndicate*, 26 April 2021. Retrieved 28 June 2021 from https://www.project-syndicate.org/commentary/eu-geopolitics-of-climate-change-by-frans-timmermans-1-and-josep-borrell-2021-04.

UNDP (2020) 'Human Development Index'. Retrieved 29 June 2021 from http://hdr.undp.org/en/content/human-development-index-hdi.

Van de Graaf, T. and Verbruggen, A. (2015) 'The oil endgame: Strategies of oil exporters in a carbon-constrained world', *Environment Science & Policy*, 54, pp. 456–462.

Van de Graaf, T. and Sovacool, B.K. (2020) *Global energy politics*. Cambridge, Oxford, Boston, MA, New York: Polity.

Van de Graaf, T., Overland, I., Scholten, D., and Westphal, K. (2020) 'The new oil? The geopolitics and international governance of hydrogen', *Energy Research and Social Science*, 70, 101667.

Von der Leyen, U. (2019). 'Josep Borrel, High Representative of the Union for Foreign Affairs and Security Policy/Vice President of the European Commission'. 1 December 2019. Mission letter. Available at: https://ec.europa.eu/commission/commissioners/sites/default/files/commissioner_mission_letters/mission-letter-josep-borrell-2019_en.pdf. (accessed 29 June 2021).

World Bank (2020a) 'Worldwide Governance Indicators'. Retrieved 29 June 2021 from https://databank.worldbank.org/source/worldwide-governance-indicators.

World Bank (2020b) 'RISE: Regulatory Indicators for Sustainable Energy 2020: Sustaining the Momentum'. Retrieved 29 June 2021 from https://openknowledge.worldbank.org/handle/10986/34937.

World Bank (2021a) 'Gas Rents as % GDP'. Retrieved 12 July 2021 from https://data.worldbank.org/indicator/NY.GDP.NGAS.RT.ZS.

World Bank (2021b) 'Oil Rents as % GDP'. Retrieved 12 July 2021 from https://data.worldbank.org/indicator/NY.GDP.PETR.RT.ZS.

World Bank (2021c) 'Coal Rents as % GDP'. Retrieved 12 July 2021 from https://data.worldbank.org/indicator/NY.GDP.COAL.RT.ZS.

World Bank (2021d) 'Fuel Exports (% of Merchandise Exports)'. Retrieved 12 July 2021 from https://data.worldbank.org/indicator/TX.VAL.FUEL.ZS.UN

Wurzel, R.K.W., Connelly, J., and Liefferink, D. (eds.) (2017) *The European Union in international climate change politics: Still taking a lead?* Abingdon: Routledge.

Youngs, R. (2014) *A new geopolitics of EU energy security*. Vanguardia Dossier. Brussels: Carnegie. Available at: https://carnegieeurope.eu/2014/09/23/new-geopolitics-of-eu-energy-security-pub-56705 (accessed 29 June 2021).

2 Nigeria

Addressing fragility away from oil?

Introduction

The Federal Republic of Nigeria (henceforth Nigeria) is a hugely diverse country with over 250 ethnic groups and a growing, youthful populace. With the largest population and the biggest economy in Sub-Saharan Africa, it is also a major force on the African continent. By 2020, its gross domestic product (GDP) rose to 25th in the world (World Bank 2021a), with this growth largely driven by revenues from oil and gas extraction.

However, the billions of dollars generated each year in oil and gas revenues have not brought prosperity to most Nigerians, with GDP per capita ranking only the 176th highest globally (CIA 2021). Unlike in some other resource-rich countries, where the wealth generated has been invested in developing key public goods, such as education, healthcare, or transport networks, Nigeria's fossil fuel industries have mainly enriched a small minority of Nigerians, as well as foreign investors. As such, Nigeria is often referred to as a case illustrating the 'resource curse', with oil and gas profits breeding corruption, fuelling conflict between ethnic groups, undermining governance and institutional frameworks, and holding back other areas of the economy. The country features low down on indexes measuring overall well-being, development, and inclusiveness (Sachs et al. 2017; UNDP 2018; WEF 2018).

The oil and gas industries have also had a devastating impact on the environment and climate. Liquid pollutants and gas flaring[1] have contaminated the oil-producing Niger Delta's rich ecosystem and posed health risks for around 20 million people (Anejionua et al. 2015). Aside from the major climate impacts of its oil and gas exports in other countries, gas flaring alone in Nigeria is estimated over the past 50 years to have emitted more carbon dioxide (CO_2) than 1,000 coal-fired power plants would in a year.[2] Nigeria is highly vulnerable to the impacts of climate change and has been ranked the 21st last country in the ND-GAIN Index measuring vulnerability and readiness (ND-GAIN 2020).

As has been done for other case studies presented in this book, this chapter too will explore the challenges and opportunities Nigeria and Nigeria-European Union (EU) relations are facing under decarbonisation in five steps. The next section first substantiates Nigeria's high dependence on the production and export of oil and gas and hence its vulnerability to price fluctuations and the phase-out of fossil fuels. It also analyses other fragility and security risks, including internal terrorist and security challenges. Subsequently, the status of Nigeria's climate policy is assessed, which leaves significant room

DOI: 10.4324/9781003183037-2

for further improvement. This is followed by the review of relevant trends and potentials, including with respect to agriculture and other low-carbon sectors. The one but last section reviews the framework and basis of EU-Nigeria relations beyond fossil fuels. The final section concludes and identifies focal points for EU foreign policy towards Nigeria under decarbonisation.

Exposure and risks

The oil price collapse from 2014 to 2016 and the devastating effects of the Covid-19 pandemic on international oil markets showed that Nigeria is highly exposed to shifts in the international oil market. With government revenues and foreign exchange earnings strongly linked to the oil and gas sector, these downturns had major implications for the entire Nigerian economy. Furthermore, as the world decarbonises, Nigerian fossil fuel assets risk losing their value. Nigeria also faces substantial other fragility risks and security threats and as such it is ranked 12th out of 179 countries in the 2021 Fragile State Index (see also Figures 2.2 and 2.3). High levels of corruption and the poor distribution of oil wealth, both throughout society and geographically across different regions, have resulted in significant socioeconomic inequality, fuelled ethnic tensions, and led to instability and violence in some areas. The country is also highly vulnerable to climate change impacts, with the Lake Chad region and coastal megacity Lagos being examples of high-risk areas.

Exposure to global decarbonisation trends

The fortunes of the Nigerian economy are closely tied to the oil price, making it vulnerable to oil market fluctuations and the phase-out of fossil fuels. Possessing the largest oil and gas reserves in Sub-Saharan Africa, Nigeria is the region's largest and the world's 12th largest oil producer (BP 2020). Nigeria's oil boom began when the oil industry was nationalised and the country joined the Organization of the Petroleum Exporting Countries (OPEC) in 1971. In 2019 production totalled about 2.1 million barrels per day (BP 2020). The marginal production cost of onshore oil is relatively low at US$15 per barrel and US$30 per barrel for deep-water extraction (Knoema 2018). While it is only the 17th largest producer of natural gas (BP 2020), Nigeria ranks sixth in the world in terms of liquefied natural gas (LNG) exports (International Gas Union 2021). Since Nigeria's natural gas reserves came on stream in 1998, the gas sector has been developed for domestic supply and foreign export, with production increasing by about 35 percent since 2011 to 49.3 billion cubic metres in 2019 (BP 2020). However, large quantities continue to be flared, or burnt off, as many oilfields lack the infrastructure to capture the gas they produce (Raval 2017).

The oil industry accounted for 7.4 percent of Nigeria's GDP in 2019 (World Bank 2021b). Although this figure is relatively low compared to some other oil-producing nations, revenues from the oil and gas sector are a vital source of foreign exchange and the largest single source of government revenues. Fossil fuels constitute about 90 percent of Nigeria's total goods and services exports (World Bank 2021i), and in 2019 fossil fuel exports generated US$50.3 billion, accounting for about 95 percent of Nigeria's foreign exchange earnings (Chatham House 2021; see also Figure 2.1). Tax revenues from the oil and gas sector provide by far the largest share of the government budget,[3] and, as such,

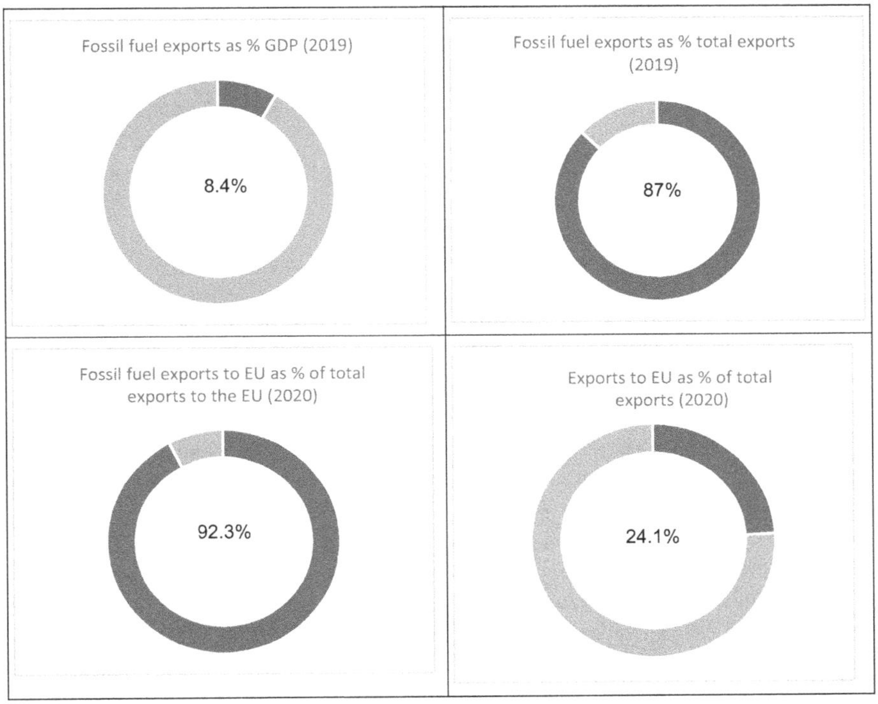

Figure 2.1 Status of fossil fuel exports of Nigeria in 2019/2020.

Sources: Data based on: https://data.worldbank.org (fossil fuel exports as a percentage of merchandise export; oil, gas, and coal as a percentage of GDP – accessed 25 July 2021) and European Commission (2021b) (exports to EU – accessed 25 July 2021).

public spending at federal and state level is dependent on production levels and international commodity markets. Non-oil government revenue comprised only 3.4 percent of GDP in 2018, one of the lowest in the world (EIA 2020). Nevertheless, the underassessment and underpayment of taxes have been a significant issue in Nigeria, and in 2017 the government realised only 53 percent of budgeted oil revenues (World Bank 2018). Furthermore, the Nigerian government supports the industry with significant subsidies. As the government began to rein in its subsidies, Nigeria has reduced government energy subsidies to less than US$1 billion in 2020 (IEA 2021).

Between 2005 and 2015, the Nigerian economy experienced impressive GDP growth of 8 percent per year on average. This was largely driven by the high oil price, with the country's exports generating the foreign currency to import most of the country's needs (Adeosun 2017). In 2019, these included US$8.8 billion worth of refined petroleum imports, 71.6 percent of which came from Europe (Chatham House 2021). However, the economy's exposure to changes in the international oil market became strikingly apparent when oil prices collapsed from more than US$100 per barrel in 2014 to under US$30 in early 2016 (BBC 2016). This broke the economy's strong upward trend, as a result of major knock-on effects on non-oil sectors dependent on imports of inputs and raw materials such as industry and services (World Bank 2017). As a result, in 2016, Nigeria experienced its first full year of recession in 25 years, with real GDP contracting by 1.6 percent (World Bank 2021c). Although still at relatively low levels, public debt as a result also rose from 17.6 percent of GDP in 2012 to 29.1 percent in 2019 (Trading Economics 2021a).

The Covid-19 crisis has further reconfirmed the risks inherent in Nigeria's continued dependence on fossil fuels. With the oil sector accounting for half of government revenue, oil revenues were 65 percent lower than expected during the first half of 2020. The Nigerian economy contracted by about 1.8 percent in 2020, even exceeding the recession of 2016 (Natural Resource Governance Institute 2020; World Bank 2021c).

Nigerian oil production has also fluctuated significantly over time. Since the mid-2000s, this has predominantly been due to militant groups sabotaging pipelines and other key infrastructure in the Niger Delta and staging kidnappings and militant takeovers of oil facilities to forward their political objectives of greater redistribution of oil wealth and local control of the oil sector (EIA 2021).

Fossil fuel reserves make up 40 percent of Nigeria's total assets (Manley et al. 2017). With proven reserves of 37 billion barrels of oil in 2019, Nigeria could maintain current oil production levels for another 48 years (BP 2020). Exploration has slowed in recent years due to the low oil price, increased security threats, and regulatory uncertainty (EIA 2021). Nonetheless, under the 450 ppm scenario,[4] estimates have suggested that by 2025 Nigeria may unnecessarily invest US$42.5 billion in capital expenditure – 22 percent of the oil sector capital expenditure for that period (Ivleva et al. 2017).

The National Petroleum Policy adopted in 2017 evaluates the potential end of the oil era and explores options for reducing oil dependency (George and Onuah 2017). One key pillar of plans to diversify the economy is to significantly expand gas production for domestic use and international export (Raval 2017; Wallace et al. 2018), given the longer-term prospects of this fuel in the global low-carbon energy transition. The country currently has discovered reserves of 5.4 trillion cubic metres of natural gas (BP 2020), which would allow it to maintain current production levels for more than 100 years. In any event, heavily investing in infrastructure to exploit these natural gas assets is likely to create stranded asset risks further down the line.

Although diversifying the economy has become a government priority, the rise in federal revenues from petroleum extraction from the 1970s was accompanied by decades of neglect of other key economic sectors, particularly agricultural production (Eigege and Cooke 2016). Prior to the oil boom, Nigeria was a world leader in the production of cocoa, palm oil and kernel, and other cash crops, with agriculture contributing over 60 percent to GDP. In 2020, this figure stood at 24 percent (World Bank 2021d), and Nigeria is losing an estimated US$10 billion in annual export opportunity from palm oil, cocoa, and other cash crops due to a decline in the production of these goods (FAO 2018). Although the declining fortunes of the agricultural sector may in part be attributable to – and corresponding to – the reduced global importance of this sector relative to others, the competitiveness of Nigerian agricultural products may also have suffered the effects of the 'Dutch disease' – an appreciating exchange rate due to the overwhelming amount of oil-based foreign currency earnings. Nigeria went from being a major exporter of agricultural products to an importer of basic foodstuffs, such as wheat (Otaha 2012; Eigege and Cook 2016).

Other fragility and security risks

While its oil and gas reserves have brought huge wealth to the country, the profits and benefits have not been equitably shared or sufficiently invested in key public services and infrastructure. Even the Nigerian economy's impressive growth in the years prior to the recent recessions did not result in significant poverty reduction or job creation. Before the pandemic, about 53 percent of the Nigerian population lived on less than US$1.90 per day[5] and more than 30 percent lived in severe multidimensional poverty (UNDP 2018). However, there are wide disparities between different regions, with those living in the north of the country much more likely to be deprived and vulnerable (AfDB 2018: 8). Decades of underinvestment in public services and infrastructure have left 59 million Nigerians without clean water (WaterAid 2018), and most without access to public transport or more than rudimentary health facilities (Holman 2017). In spite of its abundant oil and gas resources, and huge renewable energy potential, Nigeria is also locked in a 'full-blown energy crisis' (Raval 2017; see also below). The country has gradually become more unequal in terms of wealth, income, opportunities, basic services, and intergenerational equity (WEF 2018). If anything, the Covid-19 pandemic has further worsened the situation (World Bank 2021e; see also Figures 2.2 and 2.3).

Nigeria also faces significant demographic pressures. The country is estimated to have had a population of around 196 million in 2018, and, despite the birth rate falling, its population is estimated to reach 264 million in 2030 and 411 million in 2050 (IARAN 2016; UN DESA 2018). As such, around 1.8 million young Nigerians enter the workforce every year and GDP growth may have to reach into double figures for job creation to keep pace (Alemu 2015). Unemployment has been constantly rising, especially during the 2014–2016 economic downturn when unemployment increased in both rural and urban areas, almost tripling between late 2014 and the third quarter of 2017 to 18.8 percent, particularly among young people. As a result of the recession following the Covid-19 pandemic, unemployment jumped to 33.3 percent in January 2021 (Trading Economics 2021b).[6] Significant job creation is unlikely to come from the oil and gas sector, which accounts for less than 1 percent of jobs in Nigeria (PWC 2016: 23). By contrast, the agricultural sector employs over 60 percent of the labour force but is characterised by low productivity and production, with the latter increasingly impacted by climate change (Eigege and Cooke 2016).

Fragility (2021)	Sustainable	Stable	Warning	**ALERT**
Human development (2019)	Very high	High	Medium	**LOW**
Strength of governance (2019)	Very high	High	Medium	**LOW**
Climate change vulnerability (2019)	Low	Medium	**HIGH**	Very high
Sustainable energy development (2019)	Very high	**HIGH**	Medium	Low

Figure 2.2 Governance indicators (Nigeria).

Sources: Fund for Peace (2021), ND-GAIN (2020), UNDP (2020), World Bank (2021f, 2021j).

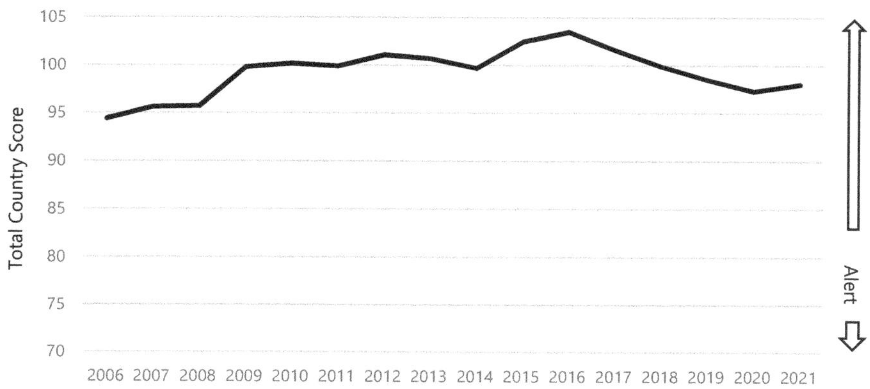

Figure 2.3 Nigeria's performance in the Fragile States Index 2006–2021

Although military rule ended in Nigeria in 1999, the first democratic transfer of power only occurred in 2015 when opposition candidate General Muhammandu Buhari was elected president. The country performs poorly on indexes measuring voice and accountability, government effectiveness, regulatory quality, rule of law, and control of corruption. Nigeria performed poorly in the 2019 World Bank Worldwide Governance Indicators for Voice and Accountability (percentile rank 35), Political Stability and Absence of Violence (5.2), Government Effectiveness (13.5), Regulatory Quality (17.8), Rule of Law (18.75), and Control of Corruption (13) (World Bank 2021f; see also Figure 2.2) and ranked 149 out of 180 countries in the Transparency International Corruption Perceptions Index 2020, scoring 25 out of a potential 100. Despite Buhari campaigning for the presidency on the promise to crack down on corruption, the country's score has slightly worsened since 2014 (27/100, rank 136/175) (Transparency International 2021).

In particular, the country has faced significant challenges in managing the unaccountable use of revenues and corruption in the oil and gas sector, with Nigeria's central bank governor warning that US$20 billion in sales revenues from the Nigerian National Petroleum Corporation (NNPC) had disappeared (NRGI 2015). Despite efforts by the government, a 2017 study by Nigeria's National Bureau of Statistics and the UN Office on Drugs and Crime showed that corruption is still rife, with an estimated 82.3 million bribes, amounting to US$1.1 billion paid out each year (Johnson 2017). This has consequences for the provision of public services and the wider economy.

Weak governance represents a significant barrier to diversifying government revenue streams beyond the oil and gas industries. Reform of Nigeria's 'cumbersome and ambiguous' tax system has long been needed (PWC 2016), with IMF research suggesting the informal economy may account for as much as 65 percent of the country's GDP, by far the highest proportion of the 37 Sub-Saharan Africa countries compared in its study (Hoffmann et al. 2015: 2; IMF 2017: 51). Increasing non-fossil-fuel trade would reduce volatility in export revenues. However, the 'overwhelming' complexity and cost of trading procedures in Nigeria and the near impossibility of controlling its 4,000-kilometre land border have created strong disincentives to trading through formal channels and resulted in a sophisticated network of unrecorded trade in non-oil sectors. It has been estimated that official government statistics may not reflect between 70 and 80 percent of Nigeria's total trade (Hoffmann et al. 2015: 2).

Nigeria also faces numerous security challenges, with development efforts regularly undermined by terrorist attacks and regional conflicts in the northeast, the Niger Delta, and the Middle Belt, each of which have different drivers and economic implications (Johnson 2017). Control of Nigeria's oil wealth has been at least a contributing factor in many of the country's most violent conflicts, notably the bloody secessionist Biafran War (1967–1970) and the security challenges in the oil-producing Niger Delta since the early 2000s, the latter fuelled by militant groups seeking greater local control and revenues from the oil industry. Secessionist sentiments have increased in the southeast (Munshi 2018). The wide disparities in income and opportunity across society and the country are another key root cause of conflict, with the security challenges in the northeast and the brutal insurgency of the Islamist sect Boko Haram driven by 'a dangerous mix of unemployment, depleting resources, economic hardship and violent conflict' (Nett and Rüttinger 2016: 10). As a result, more than 2.8 million people are currently internally displaced, and, aggravated by the Covid-19 pandemic, about 9.2 million people countrywide are food-insecure, of whom about 3.2 million live in the north-eastern states of Adamawa, Borno, and Yobe (FAO 2021; UNHCR 2021). Intensifying competition for fertile land is also contributing to the country's security challenges, particularly in the northeast and Middle Belt, where clashes between farmers and herdsmen over access to farming or grazing land and water have been frequent (Eigege and Cooke 2016). Although the administration of President Muhammadu Buhari has made efforts to calm these conflicts, both regions remain fragile, posing a risk to political and economic stability in Nigeria (Johnson 2017; Crisis Watch 2018).

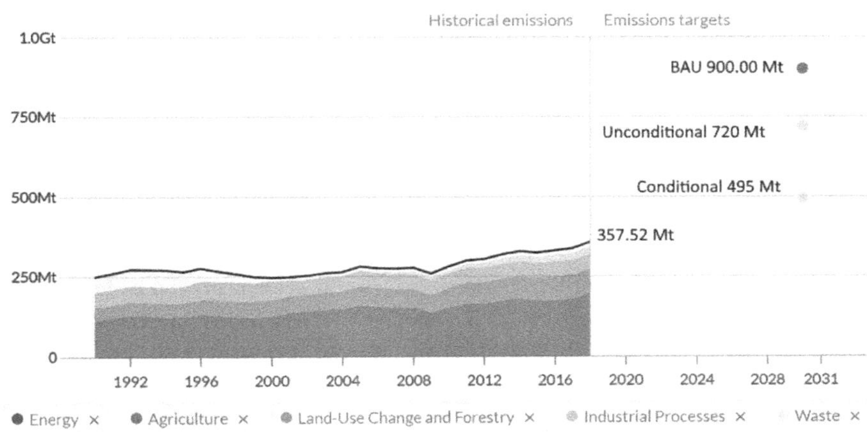

Figure 2.4 Nigeria's greenhouse gas emissions and emission targets (1990–2018)

Furthermore, Nigeria is vulnerable to the impacts of climate change and already struggling with increasing temperatures, erratic rainfall, desertification, rising sea levels, and hydrological drought (ND-GAIN 2020; Germanwatch 2021; see Figure 2.2). In the agricultural sector, these impacts have diminished crop yields and 'disrupted long-held crop rotation practices and traditions'. Food insecurity is on the rise, particularly in northern Nigeria, where agriculture is the economy's 'lifeblood' and which had thus already been disproportionally disadvantaged by the sector's decades of decline. The variability in the water levels of Lake Chad and the unfolding crisis in the region have put additional pressure. Climate change also poses significant challenges for Nigeria's rapidly growing urban centres. Nigeria's and Africa's largest city Lagos, which has a population of 21 million and 70 percent of whom live in slums, is also exposed to sea-level rise and flooding given its low-lying position on the Atlantic coast (Slaughter and Odume 2017). Furthermore, Nigeria has a long road ahead when it comes to preparing for the impacts of climate change, ranking 183 out of 192 countries in terms of its economic, governance, and social readiness (ND-GAIN 2020).

Past and present efforts to decarbonise

Nigeria is party to most major climate treaties, having ratified the United Nations Framework Convention on Climate Change (UNFCCC) in 1994, the Kyoto Protocol in 2004, the Doha Amendment in 2020, and the Paris Agreement in 2016. It is a non-Annex I Party to the UNFCCC and is hence considered a developing country in climate treaties. In its original Nationally Determined Contribution (NDC) to the Paris Agreement, Nigeria committed to a 20 percent emission reduction relative to a business-as-usual scenario by 2030, and, if international support is provided, to increase this to 45 percent. The latter would ensure that Nigeria's emissions remained almost stable up until 2030. The NDC identifies climate-smart agriculture and land use, renewable electricity generation, energy efficiency, the oil and gas sector, and transport as priority areas for reducing emissions (FRN 2015). In May 2021, Nigeria submitted an interim update of its first NDC (to be followed by a final update later on in 2021) that updated data and projections and otherwise reconfirmed the ambition of the original NDC (FRN 2021).

Nigeria accounts for less than 1 percent of total global GHG emissions (FRN 2015). Nigeria's GHG emissions amounted to 357.5 Mt CO_2e in 2018, and energy and land use, land use change, and forestry (LULUCF) are by far the highest GHG-emitting sectors (WRI 2021; see also FRN 2021; Figure 2.4). Having lost around 20 percent of its forest cover between 1990 and 2018, reducing deforestation and implementing afforestation programmes will be crucial for both adaptation and mitigation in Nigeria (World Bank 2021g).

Nigerian climate policy is still in the early stages of development, and its legislature has yet to enact a climate change law with binding emission reduction targets. However, there have been new initiatives to strengthen and mainstream climate policy and governance, which falls under the remit of the Ministry of Environment. With regard to climate change mitigation and adaptation policy, the National Climate Change Policy Response and Strategy, adopted in 2012, provides the framework and central vision for the implementation of Nigeria's NDC, while the National Policy on Climate Change, approved a year later, will form the basis for any new climate change laws. However, a climate change bill passed by the parliament in 2018 was blocked by the president in 2019, and by early 2021 a revised bill had still not been enacted (GRICCE 2021; see also FRN 2020b).

The oil price collapse and ensuing recession strengthened the political will for structural reform and diversifying the Nigerian economy beyond fossil fuels. In early 2017, the Buhari administration launched an overarching economic development plan – the Economic Recovery and Growth Plan 2017–2020. In June 2020, an Economic Sustainability Plan was adopted that aims to create 250,000 jobs and power 5 million households by 2023 through the expansion of solar power. However, policies to promote renewable energy production and use, energy efficiency, climate change adaptation, and other elements of climate policy remain to be further specified (GRICCE 2021), and the government envisions oil and gas continuing to play a major role in the Nigerian economy (FRN 2017; FRN 2020b). In this context, the response to the Covid-19 crisis has been found to lack on green spending (Global Recovery Observatory 2021).

Enacting the structural reforms to implement Nigeria's NDC and other policies aimed at diversifying and decarbonising the economy will require sustained political will and strong governance at national, state, and local levels. Among the governance concerns already addressed above, the extent to which Nigerian politicians and public servants may have vested interests in the country's fossil fuel industries is of particular concern for climate policy. While these sectors lack transparency and accountability, repeated corruption scandals and the slow pace of reform suggest significant links between public officials and the fossil fuel lobby. Furthermore, aside from revenues from these sectors forming a significant proportion of public budgets, the Nigerian government is heavily invested in the country's oil sector via the state-owned NNPC. In 2019, the NNPC was directly responsible for more than a third of Nigerian oil production, and in 2015 it sold almost half of the country's oil production (NRGI 2015). It also plays a central role in all stages of decision-making relating to the oil sector (NRGI 2018).

Although Nigeria has made 'meaningful progress' to improve the transparency of the oil and gas sector (EITI 2017), there is still an acute lack of information on revenue flows and the beneficiaries. Despite top officials' repeated commitments to tackling the poor governance of the NNPC and the sector in general, Africa's largest state-owned enterprise has yet to produce comprehensive annual reports detailing its finances or to disclose information on the earnings by its subsidiaries, the costs of its operations, and its significant spending on non-commercial activities. A particular area for concern is the licensing and contracts for oil and gas projects, as the Nigerian government 'does not regularly disclose government officials' financial interest in the extractive sector or identities of beneficial owners of extractive companies' (NRGI 2018).

Trends and potential

Nigeria is a country with huge resources for developing beyond fossil fuels, including a growing, youthful population and enormous renewable energy potential. However, to take advantage of these opportunities, it requires major investment in its low-carbon sectors and its overstretched public infrastructure and services, which are essential for promoting economic diversification and more climate-friendly and inclusive development. For example, Nigeria's erratic and unreliable power grid and the gaps in its public education system are significant barriers to development and job creation.

Financing low-carbon development

Nigeria faces a huge investment gap in areas that are essential for promoting economic diversification and more climate-friendly and inclusive development. Alongside fiscal reforms and other measures to boost public revenues and spending in areas such as education and power infrastructure, private sector investment is needed to push the development of low-carbon sectors. Foreign, diaspora, and domestic investors have generally been wary of investing in the country, due to the largely poor state of public infrastructure and services, combined with the country's endemic corruption, weak institutions, poor enforcement of regulation and laws, low skill levels, and macroeconomic uncertainty (PWC 2016: 4; Barungi et al. 2017: 8).

To encourage investor confidence, the current administration has allocated significant public spending to power and transport infrastructure, including a National Integrated Infrastructure Master Plan adopted in 2021 (Ivleva et al. 2017; PWC 2018; GRICCE 2021), and enacted reforms to improve the business environment. These resulted in Nigeria moving up 39 ranks in the World Bank's Ease of Doing Business Index from ranking 170 (out of 189) in 2015 (World Bank 2015) to ranking 131 (out of 190) in 2020 (World Bank 2020). Nigeria's growing population, including its expanding middle class[7] and cities, is likely to generate a significant increase in demand for energy and consumer goods over the coming decades (PWC 2016), and this, combined with relatively underdeveloped consumer and e-commerce markets, means Nigeria could potentially offer significant opportunities for smart investors. The vibrant digital start-up scenes in cities like Lagos have increasingly attracted foreign investment (Green 2017). Nigerians in diaspora also contributed US$21 billion to the economy in 2015, and 'local Nigerians also have billions of dollars locked in low-performing assets because people are unsure of the investor environment' (Heinrich Böll Foundation 2017: 20).

The Buhari administration also took a significant step forward in its low-carbon transition in December 2017 when it launched Africa's first Sovereign Green Bond to fill the funding gap generated by the recession and raise capital for NDC implementation, which the World Bank has estimated will require US$142 billion by 2030 (Gyamfi 2017). The bonds are being used to finance projects to protect the environment and to mitigate and adapt to the impacts of climate change, and the first issuance of 10 billion naira (around US$27.6 million) successfully financed three projects related to reforestation and promoting energy security via off-grid solar and hydropower systems. The government issued a second bond (Series II Green Bond) in 2019, funding 23 projects in five sectors (afforestation, renewable energy, transport, agriculture, and water) and has also made moves towards a third bond issuance (FRN 2020a).

Potential in key low-carbon sectors

Responsible for a large proportion of Nigeria's GHG emissions, the energy sector is an important focal area for Nigeria's low-carbon development. Only around 60 percent of Nigeria's population are connected to the electricity grid, and 80 percent of those with grid access rely on generators running on expensive imported diesel fuel to cope with frequent outages (IEA 2017). In rural areas, 75 percent of people do not have access to electricity and instead rely on wood, charcoal, manure, and crop residues for heating and cooking. In northern Nigeria, this is likely the case for 99 percent of families, and the need for firewood is a major driver of deforestation (BNEF 2017). Expanding electricity access rapidly enough to meet existing, let alone rising future demand, is a major concern (BNEF 2017). Required electricity sector investment over the next 25 years has been estimated to be as high as US$100 billion (Heinrich Böll Foundation 2017: 20).

However, the scope for improvements on both demand side and supply side is huge. With energy efficiency regulation currently underdeveloped (World Bank 2017), there are a range of options available for decreasing energy consumption. Many of these – from introducing more efficient stoves in rural areas to reducing gas flaring – offer co-benefits like reduced air pollution alongside cost-savings. Nigeria is considered to have huge renewable energy potential, particularly with regard to solar PV. While Nigeria's installed solar capacity was only around 28 MW in 2019 (IRENA 2021), it has been estimated that covering just 1 percent of Nigeria's land area with state-of-the-art polycrystalline PV modules could generate 207,000 GWh per year – ten times the total electricity production in Nigeria in 2011 (Ley et al. 2015: 85). The government has been slow to promote renewables, although the issuance of green bonds in December 2017 was a step forward (FRN 2020a). However, entrepreneurs and aid agencies are leading the way with innovative mini-grid and decentralised solar energy schemes in urban and rural areas (Green 2017).

As mentioned above, the Economic Sustainability Plan adopted in 2020 does foresee some limited investments in the expansion of renewable energy in general and solar power in particular. However, the response to the Covid-19 crisis has overall had only little emphasis on clean energy and green spending (Global Recovery Observatory 2021; GRICCE 2021).

The Nigerian agricultural sector also has major potential; however, for the last 50 years it has been cheaper to import rice from China, India, or Thailand than to source food in Nigeria. Increasing production and moving towards self-sufficiency, particularly for cereals, would reduce Nigeria's agricultural products import bill, which amounted to US$4.4 billion in 2019 (Chatham House 2021). The agricultural sector employs 70 percent of the population, with 95 percent of the country's US$90 billion farm output estimated to be produced by subsistence farmers, often with less than a hectare of land each (Wheatley 2017). As such, there is ample scope for increasing productivity and creating higher-paying jobs in processing, marketing, or distribution. Improvements to infrastructure would also significantly boost sector output. The delays caused by the country's unreliable power supply and congested transport networks and ports, combined with bureaucracy, corruption, and poor governance, mean that at present fruit and vegetables often perish before making it to market. Furthermore, most subsistence farmers do not have access to credit to buy the storage and processing equipment they need to efficiently market their produce. Investment is also needed to increase the sector's resilience to climate change impacts. Nigeria's original NDC stated that

> under a business-as-usual scenario agricultural productivity could decline between 10 to 25 percent by 2080. In some parts of the north, the decline in yield in rain fed agriculture could be as much as 50 percent. This in turn would impact GDP, reducing it by as much as 4.5 percent by 2050, even though the share of agriculture in GDP will decline from 40 to just 15 percent.
>
> (FRN 2015)

Despite major challenges, revitalising the sector could be central to Nigeria's future development, and the weakening of the Nigerian currency naira and various policy initiatives have begun to turn the situation around (Wheatley 2017). Government, private sector, and foreign donors have begun to promote innovative and scalable schemes to expand market access, provide credit to farmers, and reduce post-harvest losses (Downie 2017). Increasing production to become a net exporter of agricultural products would diversify and stabilise the economy by providing a major source of non-oil revenue and foreign currency. Improving nutrition and food security would also raise living standards and tackle some of the root causes at the heart of Nigeria's security challenges. For example, the food emergency in the northeast has served to significantly strengthen support for the insurgency of Boko Haram (Nett and Rüttinger 2016; Downie 2017: 7).

Education and skills development

Nigeria's growing and youthful population presents both challenges and opportunities for the future. With around 67 percent of the population under the age of 30, and 43 percent under the age of 15, Nigeria has the potential to follow the path of other emerging economies that have reaped a 'demographic dividend' as a result of a growing working population. However, this is contingent on the rapid expansion of the public school system, educational and vocational curriculums being suited to the future demands of the labour market, and on ample jobs being available for school leavers when they enter the workforce (PWC 2016; OBG 2017a). There are significant gaps in the Nigerian public education system at present. Net enrolment in primary education has remained below 70 percent (WEF 2019; World Bank 2021h). Overall, the Nigerian education system lacks investment and basic infrastructure and is characterised by poor teaching, a rigid curriculum, and large class sizes, which can reach 80 to 100 students in some regions. According to 2015 UNESCO data, one in three children drop out of school at primary level, only one in two children in the Nigerian public education system can read or write, and in rural areas two in three children cannot read at all (OBG 2017a). As regards higher education, the number of university spaces is currently inadequate to absorb the rising number of applicants, with the 1.8 million candidates who sat the university admissions exams in 2017 jostling for only 850,000 places (Adesulu 2017). Among those who do secure places, there is a clear preference for arts and social science subjects, as well as professional degrees in law, medicine, and engineering. The comparative lack of school leavers opting for vocational courses and university students studying science, technology, engineering, and mathematics (OBG 2017a, 2017b) may leave Nigeria with a workforce ill-prepared for low-carbon professions, such as those related to the design, maintenance, and installation of energy-efficient or renewable-energy technologies.

The Nigerian government has put in place policies to improve access to education and the public school and university systems. Although falling during the 2015 recession, spending on education rose from 2016, with 7 percent of the 2018 budget allocated to this policy area. However, the programme the government launched in 2016 to create more jobs for teachers, provide school food, and other measures has suffered from slow implementation (Barungi et al. 2017: 10–11). The number of private sector schools and universities has risen rapidly to fill the gaps in the overstretched public system, especially in urban areas (Härma 2016). The government has launched initiatives to better prepare more school and university leavers for jobs in sectors such as construction (OBG 2017b). Major investment is also required in university education, vocational centres, and technical colleges to educate and train the kinds of high-skilled workers needed by the country's growing services sector (PWC 2017). However, the government cut back expenditure on education in the wake of the Covid-19 crisis (Olufemi 2020).

Cooperation with the EU

EU foreign policy towards Nigeria is already providing support for many of the major challenges the country faces in making a stable transition to a non-oil economy and meeting the Sustainable Development Goals (SDGs). The EU has been active in supporting efforts to address key areas such as good governance and security as well as those promoting low-carbon development through various measures such as renewable energy and energy access, sustainable agriculture, entrepreneurship, and innovation. However, efforts to deepen trade in non-oil sectors via an Economic Partnership Agreement have stalled.

Political dialogue

EU cooperation with Nigeria has for more than 20 years found its legal basis in the Cotonou Agreement, a partnership agreement that since 2000 has provided the framework for cooperation with 79 developing countries in Africa, the Caribbean, and Pacific (ACP) in the fields of development cooperation, trade, and political dialogue. The Cotonou Agreement was due to expire in 2020, but the parties agreed to extend its application until November 2021. Political negotiations for a post-Cotonou agreement bore fruit in late 2020, but the new agreement still has to be signed and ratified by the parties. The new partnership agreement with African nations will aim to support the implementation of the 2030 Agenda for Sustainable Development, including on peace and stability, democracy and good governance, and economic and human development (European Commission 2020).

The EU and Nigeria began to intensify political dialogue beyond the Cotonou Agreement in 2004, and in 2009 established the 'Nigeria-EU Joint Way Forward' as a new political framework for dialogue. Agreed priority issues of mutual interest and concern include peace and security; good governance and human rights; economic development, including trade and regional integration; energy; and environmental sustainability and climate change. The core of the new framework is annual ministerial dialogues. In 2018, the EU made commitments to support Nigeria in the fight against terrorism, most notably Boko Haram, and offered support in strengthening democratic governance and institution-building in the run up to the 2019 general elections (EEAS 2018a, 2018b).

Trade relations

Nigeria and the EU maintain strong economic ties. Nigeria is the destination for around half of EU exports to the West African region and close to 70 percent of imports (EEAS 2016). Fossil fuels, and particularly oil, are the main pillar of this trading relationship (see Figure 2.1). In 2019, fossil fuels and mining products accounted for 96 percent

(€20.6 billion) of total Nigerian goods exports to the EU and 54 percent (€6.3 billion) of total EU goods exports to Nigeria (European Commission 2021a; see also Chatham House 2021). While EU crude oil imports are highly diversified, with Nigeria providing only 7.3 percent (worth US$17.7 billion) in 2019, refined oil and oil products from the EU, worth US$6.3 billion, accounted for about 72 percent of Nigerian oil imports. Agricultural products, mainly cacao and leather, are the second most important commodities exported by Nigeria to the EU, gradually increasing to reach US$2.1 billion between 2000 and 2013. However, exports of these products fell back to US$687 million in 2019 (Chatham House 2021). According to the EU statistics, the EU imported Nigerian agricultural products worth €574 million in 2019, which constitutes only about 2.7 percent of total goods imports (European Commission 2021a). Important EU exports to Nigeria include agricultural products, such as wheat, milk, and frozen fish, and chemicals and machinery and transport equipment (Chatham House 2021; European Commission 2021b).

Efforts to expand trade in non-oil sectors have stalled in recent years. As of July 2021, Nigeria was the only country among the 15 member states of the Economic Community of West African States (ECOWAS), home to 300 million people, not to have signed the Economic Partnership Agreement (EPA) with the EU (European Commission 2021c). Negotiated between 2004 and 2014, the EPA offers ECOWAS countries immediate tariff-free access to the EU in return for the phased liberalisation over a 20-year period of 75 percent of the bloc's markets to EU trade. Powerful voices in Nigerian politics, including the Manufacturers Association of Nigeria, have opposed the EPA, arguing that it would interfere with the government's strategy to promote and protect burgeoning non-oil segments of the economy. The government's position has been that Nigeria's industries – even in 20 years' time – will be too underdeveloped to benefit from more liberalised trade and that the US$8.94 billion in financial and technical assistance offered to ECOWAS member states via the EPA Development Programme to prepare for competition from European products is insufficient (Boyo 2017). Alone, Nigeria's annual infrastructure investment gap is estimated to be US$8 billion, and Nigerian officials have noted that the same financial assistance is available via the European Development Fund and EU state budgets. As all ECOWAS members have to sign up to the EPA for it to enter into force, the EPA is therefore unlikely to enter into force anytime soon, especially given that other ECOWAS member states are unwilling or unable to pressure Nigeria into signing (Vasey 2017).

Development cooperation

Due to its oil and gas exports, Nigeria has never been a country dependent on aid. Nonetheless, development assistance has an important role in Nigeria's relations with the EU (European Commission n.d.). The EU institutions allocated €562 million in official development assistance to Nigeria for the period 2014–2020, with the National Indicative Programme for Nigeria outlining the main areas of cooperation, namely sustainable energy and access to electricity, health, nutrition and resilience, and rule of law, governance, and democracy (European Commission n.d.). Since 2014, the EU allocated more than €340 million of humanitarian aid to Nigeria, of which €52 million were allocated in 2021 in the context of the Covid-19 pandemic (European Commission 2021d) and €224.5 million in support to the north-eastern Borno state in 2017 in view of the growing crisis in the Lake Chad region (European Commission 2017).

In the past, by far the largest EU member state donor to Nigeria had been the United Kingdom, which provided US$431 million in assistance in 2016 (OECD 2021), with about a third of the budget being allocated to health-related projects (UK Aid 2021). Post-Brexit, EU member state contributions are therefore likely to fall significantly, with the next largest member state donors France and Germany having provided US$43 million and US$155 million, respectively (OECD 2021).

Low-carbon development

EU efforts to promote sustainable energy included the Nigeria Energy Support Programme funded by the EU Delegation to Nigeria and the German government (GIZ 2019). The programme took a range of approaches to improve on- and off-grid access to sustainable energy in Nigeria. These included support for planning and monitoring of sector strategies, assistance to existing initiatives by public actors such as the Rural Electrification Agency, and providing its expertise on creating an enabling environment for private investment in renewable energy and energy efficiency. Initiatives at member-state level have also supported renewable energy rollout in Nigeria. For example, the German government has sought to more strongly focus the Nigerian-German Energy Partnership enacted in 2008 on sustainable energy sources in recent years (Adekoya 2019).

With regard to promoting sustainable agriculture, the EU is funding numerous projects to reduce malnutrition and support sustainable agricultural livelihoods and rural adaptation to climate change impacts. Two examples are the Produce and Sell initiative being jointly implemented by Oxfam in Nigeria and the Development Exchange Center. Targeting 300,000 people in 30,000 farming households, it aims at improving the 'resilience of small-scale farmers, fishermen and livestock owners in the face of shocks from climate change, such as changing rainfall patterns, drought and desertification' (Oxfam 2018). The EU Emergency Trust Fund for Stability and Addressing Root Causes of Irregular Migration and Displaced Persons in Africa is also funding a project to promote sustainable agriculture in Nigeria's northern Borno state. Beneficiaries will receive resources and training to help them implement best practices in fish farming, animal husbandry, and agricultural production and manage their businesses (EEAS 2019).

The EU has also provided support for improving conditions for business and agricultural investment in Nigeria. The European Investment Bank (EIB) has contributed US$20 million alongside the US$50 million from the African Development Bank to support the new Nigerian Development Bank set up in 2018. The latter is also being supported by 13 EU member-state shareholders. Given that only 5 percent of the country's 37 million entrepreneurs and small businesses can access credit from financial institutions, it aims to support Nigeria's micro, small-, and medium-scale enterprises with funding and risk-sharing facilities (EIB 2018). The EU has also provided support to Nigeria and the ECOWAS region through the West Africa Competitiveness Programme to boost economic integration, growth, and jobs with €120 million funding (ECOWAS 2018).

Conclusions

Nigeria is a country where economic development has been dependent on gas and particularly oil export revenues. Although it has grown to become the largest economy in Sub-Saharan Africa, the country is grappling with a high poverty rate, corruption and poor governance, climate change impacts, rapid urbanisation, growing youth unemployment and low productivity, due to patchy energy and transport infrastructure and low access to quality education and healthcare. The country is also facing multiple security threats. Boko Haram has remained virulent in the north, and other threats, including conflicts between herdsmen in the Middle Belt and secessionist sentiment in the southeast, have become more acute. The fallout of the Covid-19 pandemic, including a severe economic recession, has aggravated these challenges.

The importance of oil revenues for government budgets and foreign exchange earnings means that Nigeria is highly exposed to shifts in the international oil market, which is increasingly being shaped by decarbonisation trends in Europe and the world. As such, decarbonisation increases the likelihood of oil price drops that sent the Nigerian economy into recession in 2015 and in 2020. This could place further strain on oil and non-oil sectors, socioeconomic welfare, social cohesion, and political stability in a country already facing a range of complex, interrelated challenges. Nigeria's challenge is to reduce security threats, strengthen governance and channel its resource wealth – revenues from its oil and gas sector as well as from other climate-friendly sectors, such as renewable energy and sustainable agriculture – into low-carbon sectors to push development in a direction that prepares it for a decarbonised future, rather than increasing its vulnerability to it.

The EU is an important trading partner for Nigeria, but to date this trading relationship has been based mainly on oil. Efforts to increase and liberalise trade in non-oil sectors have stalled. The EU and its member states are already providing support to Nigeria in many of the areas where the country faces significant challenges – including security issues and the critical need to strengthen democratic governance. These can arguably be considered as areas most fundamental to helping Nigeria to cope with the potential negative impacts of global decarbonisation on its oil sector. However, there is plenty of scope for intensifying this cooperation. It is worth noting that the United Kingdom was the member state that provided by far the most development assistance to Nigeria and that Brexit may therefore have implications for future EU-Nigeria relations.

There are numerous entry points for the EU to support Nigeria in sustainably developing its non-oil economy and preparing for decarbonisation. While trade agreements remain off the table at present, the EU can still support Nigeria via political dialogue, development cooperation, and climate finance and assistance. Key priorities include:

- **Security situation**: Support measures to prevent further escalation of humanitarian and security crises, particularly in the northeast, Middle Belt, and Niger Delta. The roots of these crises are deep and complex. A coordinated, long-term strategy is required for the EU and its member states, to engage their military, security, development, humanitarian, and environmental agencies and institutions in dialogue and cooperation with the Nigerian authorities. With the United Kingdom having been Nigeria's most important development partner among EU member states, the

EU could continue to cooperate with the United Kingdom in its relationship with Nigeria in the wake of Brexit.

- **Good governance**: Strong governance and policies will be needed to drive the transition to a low-carbon economy and harness the country's potential. The EU and member states should continue their efforts to support democracy and free and fair elections. Capacity-building instruments such as the NDC partnership can help Nigeria achieve its climate targets.
- **Sustainable energy and agriculture**: The EU can strengthen rural regions by investing in sustainable energy access and climate-smart agriculture. It can continue to use development cooperation to support climate-smart agriculture, low-tech innovations like efficient cook stoves, and rural electrification with renewable energy in order to improve the health, living standards, and livelihoods of Nigeria's rural communities.
- **Trade**: The EU can use trade relations to support Nigeria's readiness to a decarbonised world. While encouraging adoption of the Economic Partnership Agreement, Nigeria needs reduced barriers to export non-oil goods, particularly agricultural ones, to the EU. Environmental standards and climate considerations could be inserted into trade relations to develop the long-term trade relationship beyond oil.
- **Support for investment, innovation, and education**: The EU can continue to fund initiatives that support and expand access to credit for entrepreneurs and small businesses in Nigeria, particularly in sectors supporting the country's non-oil economy like agriculture, energy efficiency, and renewable energy. Targeted schemes to help young entrepreneurs and vocational programmes for jobs with long-term prospects could also play a role in tackling youth unemployment.

Overall, Nigeria and the EU have already established a good basis for cooperation. Although efforts to deepen the trading relationship between the EU and Nigeria beyond oil and gas have stalled for now, a variety of dialogues, frameworks, and different types of assistance have been set up to provide EU expertise and funding to support Nigeria in addressing many of its most pressing security, economic, and socio-political issues and to promote and strengthen Nigeria's non-oil economy. However, given the scale of challenges and intensifying risks due to decarbonisation, climate change, and demographic change, there is significant scope for scaling up support in all areas.

Notes

1 'Gas flaring is the burning of natural gas that is associated with crude oil when it is pumped up from the ground. In petroleum-producing areas where insufficient investment was made in infrastructure to utilize natural gas, flaring is employed to dispose of this associated gas' (JINN 2010: 1).

2 Calculated using the United States Environmental Protection Agency's Greenhouse Gas Equivalencies Calculator using data from Giwa et al. (2014).

3 Nigeria is a low-taxed economy with tax to GDP estimated to be 8 percent – the second lowest in Africa and fourth lowest in the world (PWC 2016).

4 Introduced in the International Energy Agency (IEA) 2014 World Energy Outlook, the 450 ppm scenario – compatible with limiting warming to 2°C – assumes that oil demand will peak around 2020.

5 In purchasing power parity (PPP) terms.
6 Nigerian statistics count anyone working under 20 hours a week as unemployed. For more information, see https://africacheck.org/reports/nigerias-unemployment-rate-18-8-widely-tweeted/.
7 Nigeria's middle class grew from 13 percent of the population in 2003/2004 to 19 percent in 2012/2013, although this increase was lower than expected given the high increases in GDP during that period. The middle class in the country's southern states grew quicker than in northern regions (Corral Rodas et al. 2019).

References

Adekoya, F. (2019) 'Nigeria, Germany explore alternative energy solutions', *The Guardian*, 17 July 2019. Retrieved 1 November 2019 from https://guardian.ng/energy/nigeria-germany-explore-alternative-energy-solutions/.

Adesulu, D. (2017) 'UTME: 1.7m candidates jostle for 850,000 admission slots', *Vanguard*, 11 May 2017. Retrieved 10 July 2018 from https://www.vanguardngr.com/2017/05/utme-1-7m-candidates-jostle-850000-admission-slots/.

Adeosun, K. (2017) 'Resetting Nigeria's economy by boosting infrastructure', *The Financial Times*, 20 February 2017. Retrieved 10 May 2018 from https://www.ft.com/content/b644d0f8-f52b-11e6-8758-6876151821a6.

AfDB – African Development Bank (2018) 'African Economic Outlook Country Note. Nigeria'. Retrieved 11 July 2018 from https://www.afdb.org/fileadmin/uploads/afdb/Documents/Generic-Documents/country_notes/Nigeria_country_note.pdf.

Alemu, Z.G. (2015) *The Challenge of Job Creation in Nigeria*. African Development Bank. Africa Economic Brief. Available at: https://www.afdb.org/fileadmin/uploads/afdb/Documents/Publications/AEB_Vol_6_Issue_8__2015_The_Challenge_of_Job_Creation_in_Nigeria.pdf (accessed 15 February 2018).

Anejionua, O.C.D., Whyatt, J.D., Blackburn, A.G., and Price, C.S. (2015) 'Contributions of gas flaring to a global air pollution hotspot: Spatial and temporal variations, impacts and alleviation', *Atmospheric Environment*, 118, 184–193. https://doi.org/10.1016/j.atmosenv.2015.08.006.

Barungi, B., Odhiambo, O., and Asogwa, R. (2017) *African Economic Outlook. Country Notes Nigeria*. AfDB, OECD, UNDP. Available at: https://www.afdb.org/fileadmin/uploads/afdb/Documents/Publications/AEO_2017_Report_Full_English.pdf (accessed 12 February 2018).

BBC (2016) 'Oil price steadies after falling below $28 a barrel', *BBC*, 18 January 2016. Retrieved 8 April 2020 from http://www.bbc.com/news/business-35340893.

BNEF – Bloomberg New Energy Finance (2017) 'Climatescope 2017. Nigeria'. Retrieved 2 May 2018 from http://global-climatescope.org/en/country/nigeria/#/enabling-framework.

Boyo, H. (2017) 'Nigeria: EPA as "Enslavement Partnership Agreement"', *Vanguard*, 18 September 2017. Retrieved 6 December 2018 from https://allafrica.com/stories/201709190873.html.

BP (2020) *BP Statistical Review of World Energy*. Available at: https://www.bp.com/content/dam/bp/business-sites/en/global/corporate/pdfs/energy-economics/statistical-review/bp-stats-review-2021-full-report.pdf (accessed 5 July 2021).

Chatham House (2021) 'Resource Trade. Earth'. Retrieved 26 July 2019 from http://resourcetrade.earth/.

CIA – Central Intelligence Agency (2021) 'The World Factbook. Country Comparisons. Real GDP Per Capita'. Retrieved 12 July 2021 from https://www.cia.gov/the-world-factbook/field/real-gdp-per-capita/country-comparison.

Climate Action (2018) 'Nigeria Plans to Go Big on Green Bonds This Year'. 5 March 2018. Retrieved 2 May 2018 from http://www.climateactionprogramme.org/news/nigeria-plans-to-go-big-on-green-bonds-this-year.

Corral Rodas, P.A., Molini, V., and Oseni, G. (2019) 'No condition is permanent: Middle class in Nigeria in the last decade', *The Journal of Development Studies*, 55(2), pp. 294–310. https://doi.org/10.1080/00220388.2017.1366453.

Crisis Watch (2018) 'Crisis Watch Conflict Tracker for April/May 2018. Nigeria'. Retrieved 4 May 2018 from https://www.crisisgroup.org/africa/west-africa/nigeria.

Downie, R. (2017) *Growing the Agricultural Sector in Nigeria*. Centre for Statistics and International Studies. Available at: https://csis-prod.s3.amazonaws.com/s3fs-public/publication/170317_Downie_AgricultureSectorNigeria_Web.pdf (accessed 7 July 2018).

ECOWAS (2018) 'ECOWAS, EU, WAEMU Hold Strategic Steering Committee Meeting for West Africa Competitiveness Programme', 10 October 2018. Retrieved 7 July 2021 from http://www.ecowas.int/ecowas-eu-waemu-hold-strategic-steering-committee-meeting-for-west-africa-competitiveness-programme/.

EEAS (2016) 'Nigeria and the EU'. Retrieved 2 April 2018 from https://eeas.europa.eu/delegations/nigeria/1621/nigeria-and-eu_en.

EEAS (2018a) 'Nigeria: EU Investing Heavily in Security and Human Development'. Retrieved 2 April 2018 from https://eeas.europa.eu/headquarters/headquarters-homepage/38397/nigeria-eu-investing-heavily-security-and-human-development_en.

EEAS (2018b) 'The European Union Together with Ten Partners Launched a €26.5 Million Program in Support of Democratic Governance in Nigeria'. Retrieved 2 April 2018 from https://eeas.europa.eu/delegations/nigeria/39246/european-union-together-ten-partners-launched-%E2%82%AC265-million-program-support-democratic_en.

EEAS (2019) 'Restoring and Promoting Sustainable Agriculture Based Livelihoods for Food Security, Employment, and Nutrition Improvement in Borno State'. Retrieved 7 July 2021 from https://eeas.europa.eu/delegations/nigeria/56284/restoring-and-promoting-sustainable-agriculture-based-livelihoods-food-security-employment-and_en.

EIA (2020) 'Country Analysis Executive Summary: Nigeria'. Retrieved 5 July 2021 from https://www.eia.gov/international/content/analysis/countries_long/Nigeria/NigeriaCAXS_2020.pdf.

EIA (2021) 'Background Reference: Nigeria'. Retrieved 5 July 2021 from https://www.eia.gov/international/analysis/country/NGA/background.

EIB – European Investment Bank (2018) 'EIB and African Development Bank to Support Private Sector Investment in Nigeria with Development Bank of Nigeria Backing'. Retrieved 14 January 2019 from http://www.eib.org/en/infocentre/press/releases/all/2018/2018-093-eib-and-african-development-bank-to-support-private-sector-investment-in-nigeria-with-development-bank-of-nigeria-backing.htm.

Eigege, J. and Cooke, J. (2016) *Tracing the Roots of Nigeria's Agricultural Decline*. Washington, D.C.: Centre for Statistics and International Studies. Retrieved 7 February 2018 from: https://csis-prod.s3.amazonaws.com/s3fs-public/publication/160505_Eigege_NigeriasAgriculturalDecline_Web.pdf.

EITI – Extractive Industries Transparency Initiative (2017) 'Nigeria'. Retrieved 6 July 2021 from https://eiti.org/news/nigeria-eiti-recognises-progress-in-addressing-natural-resource-governance.

European Commission (n.d.) 'Nigeria'. Retrieved 7 July 2021 from https://ec.europa.eu/international-partnerships/where-we-work/nigeria_en.

European Commission (2017) 'EU Announces €143 Million Support Package for the Crisis in North East Nigeria'. Retrieved 2 April 2017 from http://europa.eu/rapid/press-release_IP-17-1645_en.htm.

European Commission (2020) 'Post-Cotonou: Negotiators Reach a Political Deal on a New EU/Africa-Caribbean-Pacific Partnership Agreement'. Retrieved 7 July 2021 from https://ec.europa.eu/commission/presscorner/detail/en/ip_20_2291.

European Commission (2021a) 'Trade Statistics. Overview. Nigeria'. Retrieved 7 July 2021 from https://webgate.ec.europa.eu/isdb_results/factsheets/country/overview_nigeria_en.pdf.

European Commission (2021b) 'European Union. Trade in Goods with Nigeria'. Retrieved 7 July 2021 from https://webgate.ec.europa.eu/isdb_results/factsheets/country/details_nigeria_en.pdf.

European Commission (2021c) 'Overview of Economic Partnership Agreements. Updated March 2021'. Retrieved 7 July 2021 from https://trade.ec.europa.eu/doclib/docs/2009/september/tradoc_144912.pdf.

European Commission (2021d) 'European Civil Protection and Humanitarian Aid Operations. Nigeria'. Retrieved 7 July 2021 from https://ec.europa.eu/echo/where/africa/nigeria_en.

FAO – Food and Agriculture Organisation of the United Nations (2018) 'Nigeria at a Glance'. Retrieved 2 February 2018 from http://www.fao.org/nigeria/fao-in-nigeria/nigeria-at-a-glance/en/.

FAO (2021) 'Northeastern Nigeria – Response Overview (April 2021)'. Retrieved 6 July 2021 from http://www.fao.org/emergencies/resources/documents/resources-detail/en/c/1398444/.

FRN – Federal Republic of Nigeria (2015) 'Nigeria's Intended Nationally Determined Contribution'. Retrieved 12 February 2018 from http://www4.unfccc.int/ndcregistry/PublishedDocuments/Nigeria%20First/Approved%20Nigeria's%20INDC_271115.pdf.

FRN (2017) 'Economic Recovery & Growth Plan 2017–2020'. Retrieved 14 October 2018 from https://yourbudgit.com/wp-content/uploads/2017/03/Economic-Recovery-Growth-Plan-2017-2020.pdf.

FRN (2020a) 'Department of Climate Change. Green Bonds'. Retrieved 6 July 2021 from https://climatechange.gov.ng/2020/09/21/brief-on-green-bonds/.

FRN (2020b) 'Third National Communication (TNC) of the Federal Republic of Nigeria under the United Nations Framework Convention on Climate Change (UNFCCC)', March 2020. Available at: https://www4.unfccc.int/sites/SubmissionsStaging/NationalReports/Documents/187563_Nigeria-NC3-1-TNC%20NIGERIA%20-%2018-04-2020%20-%20FINAL.pdf (accessed 23 July 2021).

FRN (2021) 'Nigeria's Nationally Determined Contribution - Update'. Retrieved 22 July 2021 from https://www4.unfccc.int/sites/ndcstaging/PublishedDocuments/Nigeria%20First/NDC%20INTERIM%20REPORT%20SUBMISSION%20-%20NIGERIA.pdf.

Fund for Peace (2021) 'Fragile States Index 2021'. Retrieved 26 July 2021 from https://fragilestatesindex.org/.

George, L. and Onuah, F. (2017) 'Nigeria's aims to diversify as "era of oil booms' ends"', *Reuters*, 26 July 2017. Retrieved 16 July 2018 at https://www.reuters.com/article/us-nigeria-oil-idUSKBN1AB29R.

Germanwatch (2021) 'Global Climate Risk Index 2021'. Retrieved 17 June 2021 from https://germanwatch.org/en/19777.

Giwa, S.O., Adama, O.O., and Akinyemi, O.O. (2014) 'Baseline black carbon emissions for gas flaring in the Niger Delta region of Nigeria', *Journal of Natural Gas Science and Engineering*, 20, pp. 373–379. https://doi.org/10.1016/j.jngse.2014.07.026.

GIZ (2019) 'Nigerian Energy Support Programme II'. Retrieved 7 July 2021 from https://www.giz.de/en/downloads/NESP%20II_Factsheet.pdf.

Global Recovery Observatory (2021) 'Nigeria'. Retrieved 23 July 2021 from https://recovery.smithschool.ox.ac.uk/tracking/.

Green, M. (2017) 'Diesel-dependent Nigeria looks up to the sun', *The Financial Times*, 7 November 2017. Retrieved 2 May 2018 from https://www.ft.com/content/13e73f98-a512-11e7-8d56-98a09be71849.

GRICCE – Grantham Research Institute on Climate Change and the Environment (2021) 'Nigeria'. Retrieved 23 July 2021 from https://climate-laws.org/geographies/nigeria.

Gyamfi, C.C. (2017) 'Nigeria needs $140b for climate change commitments, says World Bank', *The Guardian*, 14 February 2017. Retrieved 15 March 2018 from https://guardian.ng/news/nigeria-needs-140b-for-climate-change-commitments-says-world-bank/.

Härma, J. (2016) 'Why low-fee private schools are struggling to take root in rural Nigeria', *The Conversation*, 20 July 2018. Retrieved 13 July 2018 from https://theconversation.com/why-low-fee-private-schools-are-struggling-to-take-root-in-rural-nigeria-62664.

Heinrich Böll Foundation (2017) 'Perspectives. Political Analyses and Commentary. Africa: Putting People Back into Infrastructure'. Issue 2 June 2017. Retrieved 12 February 2018 from https://za.boell.org/sites/default/files/perspectives_june17_web.pdf.

Hoffmann, L.K., Melly, P., and Donnelly, E. (2015) 'Nigeria's Booming Borders: The Drivers and Consequences of Unrecorded Trade in Nigeria'. Chatham House. Retrieved 12 February 2018 from https://www.chathamhouse.org/sites/files/chathamhouse/events/2015-12-15-nigeria-booming-borders-meeting-summary_0.pdf.

Holman, M. (2017) 'Mixed report card for Nigeria's President Buhari', *The Financial Times*, 7 November 2018. Retrieved 15 May 2018 from https://www.ft.com/content/39f614a4-9188-11e7-83ab-f4624cccbabe.

IARAN – Inter-Agency Regional Analysts Network (2016) *Nigeria Trend Analysis: Challenges and Opportunities of the Demographic Shift and Its Link to Stunting. An Outlook to 2030*. ACF International and Institut de Relations Internationales et Stratégiques (IRIS). Available at: http://www.iris-france.org/wp-content/uploads/2016/09/NGA-Trend-Analysis-Demography-Stunting.pdf (accessed 8 February 2018).

IEA (2017) 'Energy Access Outlook 2017. From Poverty to Prosperity'. Retrieved 6 July 2021 from https://www.oecd.org/publications/energy-access-outlook-2017-9789264285569-en.htm.

IEA (2021) 'Energy Subsidies'. Retrieved 5 July 2021 from https://www.iea.org/topics/energy-subsidies.

IMF – International Monetary Fund (2017) 'Regional Economic Outlook. Sub-Saharan Africa. Restarting the Growth Engine'. Retrieved 19 February 2018 from https://www.imf.org/~/media/Files/Publications/REO/AFR/2017/May/pdf/sreo0517.ashx?la=en.

International Gas Union (2021) *IGU 2021 World LNG Report*. Available at: https://www.igu.org/resources/world-lng-report-2021/ (accessed 5 July 2021).

IRENA – International Renewable Energy Agency (2021) 'Energy Profile Nigeria'. Retrieved 6 July 2021 from https://www.irena.org/IRENADocuments/Statistical_Profiles/Africa/Nigeria_Africa_RE_SP.pdf.

Ivleva, D., Wolters, S., Scholl, C., Schlösser, T., and Schultze, K.R. (2017) *From Riches to Rags? Stranded Assets and the Governance Implications for the Fossil Fuel Sector. Discussion Paper*. Berlin: GIZ. Available at: https://www.bmz.de/rue/includes/downloads/GIZ_2017_From_Riches_to_Rags__Stranded_Assets_and_the_Governance_Implications_for_the_Fossil_Fuel_Sector.pdf (accessed 12 February 2018).

JINN – Justice in Nigeria Now! (2010) 'Gas Flaring in Nigeria: An Overview'. Retrieved 2 May 2018 from http://justiceinnigerianow.org/gas-flaring.

Johnson, S. (2017) 'Nigeria watchers crib sheet: Economy, militancy and Buhari', *The Financial Times*, 7 November 2018. Retrieved 4 May 2018 from https://www.ft.com/content/37597ea2-9188-11e7-83ab-f4624cccbabe.

Knoema (2018) 'Cost of Oil Production by Country'. Retrieved 4 December 2018 from https://knoema.de/vyronoe/cost-of-oil-production-by-country.

Ley, K., Gaines, J., and Ghatikar, A. (2015) The Nigerian Energy Sector. An Overview with a Special Emphasis on Renewable Energy, Energy Efficiency and Rural Electrification'. 2nd Edition. Eschborn: Deutsche Gesellschaft für Internationale Zusammenarbeit (GIZ) GmbH. Available at: https://www.giz.de/en/downloads/giz2015-en-nigerian-energy-sector.pdf (accessed 15 July 2021).

Manley, D., Cust, J., and Cecchinato, G. (2017) *Stranded Nations? The Climate Policy Implications for Fossil Fuel-Rich Developing Countries.* OxCarre Policy Paper 34. Oxford Centre for the Analysis of Resource Rich Economies. Available at: https://www.oxcarre.ox.ac.uk/images/stories/papers/PolicyPapers/oxcarrepp201634.pdf (accessed 12 February 2018).

Munshi, N. (2018) 'Clashes add to tension as election draws near', *The Financial Times.* 21 November 2018. Retrieved 15 July 2021 from https://www.ft.com/content/445c52bec71f-11e8-86e6-19f5b7134d1c.

Natural Resource Governance Institute (2020) *Nigeria: Updated Assessment of the Impact of the Coronavirus Pandemic on the Extractive Sector and Resource Governance*, Briefing, 13 November 2020. Retrieved 15 July 2021 from https://resourcegovernance.org/analysis-tools/publications/nigeria-updated-assessment-coronavirus-extractive.

ND-GAIN – Notre Dame Global Adaptation Initiative (2020) 'ND-GAIN Country Index'. South Bend, IN: University of Notre Dame. Retrieved 5 July 2021 from https://gain.nd.edu/our-work/country-index/.

Nett, K. and Rüttinger, L. (2016) Insurgency, Terrorism and Organised Crime in a Warming Climate. Analysing the Links between Climate Change and Non-State Armed Groups. Berlin: adelphi. Available at: https://climate-diplomacy.org/sites/default/files/2020-10/CD%20Report_Insurgency_170724_web.pdf (accessed 15 July 2021).

NRGI – National Resource Governance Institute (2015) 'Inside NNPC Oil Sales: A Case for Reform in Nigeria'. Retrieved 9 May 2018 from https://resourcegovernance.org/analysis-tools/publications/inside-nnpc-oil-sales-case-reform-nigeria.

NRGI (2018) 'Resource Governance Index. Nigeria'. Retrieved 6 July 2018 from http://www.resourcegovernanceindex.org/country-profiles/NGA/oil-gas.

OBG – Oxford Business Group (2017a) 'Nigeria's Young Population Necessitates Improvements to the Education System'. Retrieved 20 February 2018 from https://oxfordbusinessgroup.com/overview/learning-curve-efforts-under-way-educate-line-market-demands.

OBG (2017b) 'Improved Vocational Training Aims to Meet Nigeria's Labour Market Needs. Retrieved 12 July 2018 from https://oxfordbusinessgroup.com/analysis/extra-credit-vocational-training-be-key-meeting-labour-market-needs.

OECD – Organization for Economic Cooperation and Development (2021) 'Query Wizard for International Development Statistics'. Retrieved 7 July 2021 from https://stats.oecd.org/qwids/.

Olufemi, A. (2020) 'Buhari's 2021 budget share for education is Nigeria's lowest in 10 years'. *Premium Times*, 24 October 2020. Retrieved 23 July 2021 from https://www.premiumtimesng.com/news/headlines/422829-buharis-2021-budget-share-for-education-is-nigerias-lowest-in-10-years.html.

Otaha, J.I. (2012) 'Dutch disease and Nigeria oil economy', *African Research Review*, 6(1), pp. 82–90. https://doi.org/10.4314/afrrev.v6i1.7.

OXFAM (2018) 'EU, Oxfam and DEC Partner Taraba State Government on Food Security and Rural Development'. Retrieved 14 January 2019 from https://nigeria.oxfam.org/press_release/eu-oxfam-and-dec-partner-taraba-state-government-food-security-and-rural-development-0.

PWC – PricewaterhouseCoopers and The Lagos Chamber of Commerce and Industry (2016) *Nigeria: Looking Beyond Oil.* Available at: https://www.pwc.com/ng/en/assets/pdf/nigeria-looking-beyond-oil-report.pdf (accessed 12 February 2018).

PWC (2017) *Structural Transformation and Jobless Growth in Nigeria.* Available at: https://www.pwc.com/ng/en/assets/pdf/structural-transformation-jobless-growth-Nigeria.pdf (accessed 28 August 2018).

PWC (2018) 'Nigeria's Budget Highlights 2018'. Retrieved 2 May 2018 from https://www.pwc.com/ng/en/assets/pdf/pwc-2018-nigerias-budget-highlight.pdf.

Raval, A. (2017) 'Nigerian minister calls for urgency in dash for gas', *The Financial Times*, 7 November 2017. Retrieved 21 March 2018 from https://www.ft.com/content/76926c1e-9a39-11e7-8c5c-c8d8fa6961bb.

Sachs, J., Schmidt-Traub, G., Kroll, C., Durand-Delacre, D., and Teksoz, K. (2017) *SDG Index and Dashboards Report 2017*. New York: Bertelsmann Stiftung and Sustainable Development Solutions Network (SDSN). Available at: https://s3.amazonaws.com/sustainabledevelopment.report/2017/2017_sdg_index_and_dashboards_report.pdf (accessed 15 July 2021).

Slaughter, A. and Odume, N. (2017) 'Why flooding in Nigeria is an increasingly serious problem', *The Conversation*, 15 August 2017. Retrieved 2 May 2018 from https://theconversation.com/why-flooding-in-nigeria-is-an-increasingly-serious-problem-82272.

Trading Economics (2021a) 'Nigeria Government Debt to GDP'. Retrieved 5 July 2021 from https://tradingeconomics.com/nigeria/government-debt-to-gdp.

Trading Economics (2021b) 'Nigeria Unemployment Rate'. Retrieved 5 July 2021 from https://tradingeconomics.com/nigeria/unemployment-rate.

Transparency International (2021) 'Corruption Perceptions Index 2020'. Retrieved 4 May 2018 from https://www.transparency.org/en/cpi/2020/index/nzl.

UK Aid (2021) 'Nigeria'. Retrieved 7 July 2021 from https://devtracker.fcdo.gov.uk/countries/NG.

UN DESA – United Nations, Department of Economic and Social Affairs, Population Division (2018) 'World Population Prospects: The 2018 Revision'. Retrieved 5 July 2021 from https://esa.un.org/unpd/wup/.

UNDP – United Nations Development Programme (2018) '2018 Statistical Update: Human Development Indices and Indicators'. Retrieved 5 July 2021 from http://hdr.undp.org/en/global-reports.

UNDP (2020) 'Human Development Report'. Available at: http://hdr.undp.org/en/countries (accessed 26 July 2021).

UNHCR – United Nations High Commissioner for Refugees (2021) 'Operational Data Portal. Refugee Situations. Nigeria'. Retrieved 6 July 2021 from https://data2.unhcr.org/en/country/nga.

Vasey, C. (2017) 'Bloc to bloc: Why ECOWAS cannot sign the EU's EPA', *Global Risk Insights*, 24 March 2017. Retrieved 6 December 2018 from https://globalriskinsights.com/2017/03/bloc-to-bloc-why-ecowas-cannot-sign-the-eus-epa/.

Wallace, P., Burkhardt, P., and Bala-Gbogbo, E. (2018) 'Africa oil giant needs $12 billion to avoid missing LNG boat', *Bloomberg*, 7 June 2018. Retrieved 22 April 2020 from https://www.bloomberg.com/news/articles/2018-06-06/african-oil-giant-needs-12-billion-to-avoid-missing-lng-boat.

WaterAid (2018) 'Nigeria'. Retrieved 15 May 2018 from https://www.wateraid.org/uk/where-we-work/nigeria.

WEF – World Economic Forum (2018) *The Inclusive Development Index 2018. Summary and Data Highlights*. Available at: http://www3.weforum.org/docs/WEF_Forum_IncGrwth_2018.pdf (accessed 12 February 2018).

WEF (2019) *The Global Competitiveness Report 2019*. Cologny/Geneva: World Economic Forum. Available at: http://www3.weforum.org/docs/WEF_TheGlobalCompetitivenessReport2019.pdf (accessed 6 July 2021).

Wheatley, J. (2017) 'Nigeria's "great farm" model bears fruit in time of high insecurity', *The Financial times*, 7 November 2017. Retrieved 20 March 2018 from https://www.ft.com/content/a32b0c58-a75a-11e7-ab66-21cc87a2edde.

World Bank (2015) 'Doing Business Economy Profile 2015: Nigeria'. Retrieved 6 July 2021 from https://openknowledge.worldbank.org/handle/10986/21188?locale-attribute=es.

World Bank (2017) 'Nigeria Bi-Annual Economic Update, April 2017: Fragile Recovery'. Washington, D.C.: World Bank. Retrieved 12 February 2018 from https://openknowledge.worldbank.

org/bitstream/handle/10986/26658/114996-WP-P163291-PUBLIC-NEUNoFinalfrom Publisher.pdf?sequence=1&isAllowed=y.

World Bank (2018) 'Nigeria Bi-Annual Economic Update, April 2018: Connecting to Compete'. Washington, D.C.: World Bank. Retrieved 14 May 2018 from http://documents.worldbank.org/curated/en/769551524576691390/pdf/WP-NigeriaBiannualEconomicUpdateApril FinalVersion-PUBLIC.pdf.

World Bank (2020) 'Ease of Doing Business in Nigeria'. Retrieved 6 July 2021 from https://www.doingbusiness.org/en/data/exploreeconomies/nigeria#.

World Bank (2021a) 'Gross Domestic Product 2020'. Retrieved 5 July 2021 from http://databank.worldbank.org/data/download/GDP.pdf.

World Bank (2021b) 'Oil Rents (as % of GDP)'. Nigeria. Retrieved 5 July 2021 from https://data.worldbank.org/indicator/NY.GDP.PETR.RT.ZS?locations=NG.

World Bank (2021c) 'GDP Growth (Annual%) – Nigeria'. Retrieved 20 July 2021 from https://data.worldbank.org/indicator/NY.GDP.MKTP.KD.ZG?locations=NG.

World Bank (2021d) 'Agriculture, Forestry, and Fishing, Value Added (% of GDP) – Nigeria'. Retrieved 5 July 2021 from https://data.worldbank.org/indicator/NV.AGR.TOTL.ZS?locations=NG.

World Bank (2021e) *Nigeria Development Update, June 2021: Resilience through Reforms.* Washington, D.C.: World Bank. Available at: https://openknowledge.worldbank.org/handle/10986/35786 (accessed 23 July 2021).

World Bank (2021f) 'Worldwide Governance Indicators 2019'. Retrieved 26 July 2021 from http://info.worldbank.org/governance/wgi/#reports.

World Bank (2021g) 'Forest Area (% of Land Area)'. Retrieved 20 July 2021 from https://data.worldbank.org/indicator/AG.LND.FRST.ZS?locations=NG&view=chart.

World Bank (2021h) 'School Enrolment, Primary (% Net) – Nigeria'. Retrieved 6 July 2021 from https://data.worldbank.org/indicator/SE.PRM.NENR?locations=NG.

World Bank (2021i) 'Fuel Exports (% of Merchandise Exports) – Nigeria'. Retrieved 6 July 2021 from https://data.worldbank.org/indicator/TX.VAL.FUEL.ZS.UN?locations=NG.

World Bank (2021j) *Regulatory Indicators for Sustainable Energy (RISE).* Retrieved 26 July 2021 from https://rise.esmap.org.

WRI – World Resources Institute (2021) 'Climate Watch: Nigeria: Greenhouse Gas Emissions and Emissions Targets'. Retrieved 12 July 2011 from https://www.climatewatchdata.org/countries/NGA.

3 Indonesia

Building a partnership beyond fossil fuels and palm oil

Introduction

The Republic of Indonesia (henceforth Indonesia) is the world's largest archipelagic nation, with 13,466 islands, 922 of which are permanently inhabited (CIA 2021). It is also the world's fourth largest in terms of inhabitants, with the population due to hit 300 million in the early 2030s (UN DESA 2019). The capital Jakarta is located on the densely populated island of Java, which although only around the size of England is home to 145 million people. Indonesia is urbanising faster than other countries in the region, with 57.3 percent living in cities in 2021 (CIA 2021) and 73 percent projected to do so by 2050 (UN DESA 2018).

Indonesia is the world's third largest democracy and largest Muslim-majority nation, with 87 percent of the population practicing Islam (CIA 2021). Re-elected in April 2019, its seventh president Joko Widodo, also known as Jokowi, has pledged to protect Indonesia's tradition of pluralism and moderate Islam in the officially secular country (Rizki and Cahya 2018). His relatively high approval ratings are underpinned by the sustained strength of the Indonesian economy over the past two decades since the Asian financial crisis. The Indonesian economy is by far the largest in Southeast Asia and the 16th largest in the world (World Bank 2021a). Accordingly, Indonesia is the seat of and a leading voice in the Association of Southeast Asian Nations (ASEAN) and a member of the G20.

Prior to the Covid-19 pandemic, Indonesia's gross domestic product (GDP) growth averaged around 5 percent per year, and had been accompanied by significant increases in consumption and living standards, with GDP per capita increasing 108 percent from 2000 to 2019[1] and the poverty rate falling from 35 to 2.7 percent over the same period.[2] Before the pandemic, unemployment was relatively low at 3.6 percent in 2019 (World Bank 2021d), although in the same year youth unemployment stood at 20.5 percent (World Bank 2021e). However, despite these gains, there was still significant wealth inequality, and around 40 percent of the population lived on an income only just above the poverty line. Furthermore, rapid economic development and urbanisation have led to intensifying environmental problems, including large-scale deforestation, overexploitation of marine resources, air pollution, traffic congestion, unsustainable waste management, and water pollution (CIA 2021). As a result of the Covid-19 pandemic, Indonesia' economy entered a recession for the first time in two decades in 2020 (with negative consequences for the other aforementioned indicators).

DOI: 10.4324/9781003183037-3

Following the approach of the case studies, this chapter explores the challenges and opportunities Indonesia and Indonesia-European Union (EU) relations are facing under decarbonisation in five steps. The next section first discusses Indonesia's dependence on the production and export of fossil fuels and palm oil as well as its vulnerability to related price fluctuations and global decarbonisation. It also analyses other fragility and security risks, including terrorism and climate change impacts. Subsequently, the status of Indonesia's climate policy is assessed, which leaves significant room for further improvement. This is followed by the review of relevant trends and potentials, including regarding efforts to further reduce fossil fuel subsidies and develop low-carbon sectors (energy efficiency and renewable energy). The one but last section reviews the framework and basis of EU-Indonesia relations beyond fossil fuels and palm oil. The final section concludes and identifies focal points for EU foreign policy towards Indonesia under decarbonisation.

Exposure and risks

Exposure to global decarbonisation trends

Indonesia's economic growth has been underpinned by the huge variety of natural resources at its disposal. After the Amazon, it is the world's most heavily forested region, and it also has rich deposits of hydrocarbons and minerals. Many of these natural assets have also been developed for export, and the country is the second largest exporter of coal after Australia and the world's largest exporter of palm oil – the latter accounting for a large proportion of its trade with the EU. Indonesia has diversified its export base beyond fossil fuels over the past 15 years but has become increasingly dependent on coal to meet rising domestic energy demands.

Indonesia's economic growth has been underpinned by the exploitation of its abundant hydrocarbons and rich forests. Indonesia is a major producer of fossil fuels and in 2019 was the world's 2nd, 12th, and 24th largest producer of coal, natural gas, and oil, respectively (BP 2020). Coal production more than doubled from 2009 to 2019 to about 360 million tonnes of oil equivalent. Oil production has been on an overall downward trend since the 1990s and reached 781,000 barrels per day in 2019 (BP 2020). Indonesia withdrew from the Organization of the Petroleum Exporting Countries (OPEC) in 2016 so as not to have to comply with the group's production cuts (Singgih 2017). Natural gas production has also fallen, decreasing from 78 billion cubic metres in 2009 to 67.5 billion cubic metres in 2019 (BP 2020).

In 2019, Indonesia had proven coal reserves of almost 40 billion tonnes, the sixth largest in the world. Reserves could support current production rates for another 65 years (BP 2020). In 2019, Indonesia had proven oil reserves of around 2.5 billion barrels, which would be exhausted within the next decade if production continued at current levels (BP 2020). By contrast, estimated natural gas reserves of about 1.4 trillion cubic metres would support current production levels for more than 20 years (BP 2020).

Despite enormous potential and ambitious policies to harness renewable forms of energy, coal remains the mainstay of the Indonesian electricity mix, providing 59 percent of energy generation in 2019 (IEA 2021c). With reserves that are plentiful and easy to extract and transport, coal is less expensive to produce than lower-emission fuels such as natural gas and benefits from a more favourable policy environment than increasingly cost-competitive renewables (PWC 2018). Over the past decade, the consumption of oil and natural gas has risen only gradually, while coal consumption has surged, increasing by 145 percent between 2009 and 2019 (BP 2020). The proportion of oil in the power generation mix has decreased by 80 percent since 2010, from about 20 percent of electricity generation in 2010 to about 4 percent in 2019, in line with government policies to reduce petroleum imports (PWC 2017; IEA 2021c).

Although overall per capita electricity and fuel consumption remains low in Indonesia compared to other Southeast Asian countries, demand for energy is increasing rapidly. As the country's GDP almost tripled between 2000 and 2019 in constant 2010 US$, electricity demand increased by 233 percent (IEA 2021b; World Bank 2021j). The rising incomes and energy needs of its growing and increasingly urban population, as well as government plans to ensure almost universal energy access by 2025, mean that the country's electricity demand is projected to triple by 2030 (IRENA 2017). The upsurge in energy use is expected to be highest in industry and transport, with demand growing by more than half in these sectors over the same period. The country's industrialisation continues apace, and around 75 million new scooters and motorcycles and a million new cars hit Indonesia's roads every year, adding to domestic oil demand and the heavy air pollution in its densely populated cities (IRENA 2017). A growing number of household consumers are also investing in appliances like air-conditioners, fans, and cookers (PWC 2017).

Steep falls in the oil price in the 1980s prompted government action to diversify the Indonesian economy and export base (Elias and Noone 2011; World Bank 2021g). As a result, the economy has become more balanced. Exports and taxation of fossil fuel extraction have remained a significant source of government revenues (Ivleva et al. 2017). While coal has become increasingly important, the contribution of the oil and gas sector to state revenues fell from around 20 percent in 2004 to 14 percent in 2014 and then further to 3 percent in 2016 due to the oil price slump. Initiatives to attract major new investment to revive these industries have largely been unsuccessful (Ariffin 2018).

Goods and services export revenues contributed 18.5 percent of Indonesia's GDP in 2019 (World Bank 2021f). In 2017, commodity exports made up 58 percent of total Indonesian merchandise exports, of which agricultural commodities (including palm oil) and fuels made 29 and 22 percent, respectively (UNCTAD 2019, see also World Bank 2021g). Coal is Indonesia's most valuable export commodity, and coal exports generated US$26.3 billion in 2019, making the country the world's second largest exporter of coal after Australia (Chatham House 2021). In 2019, over half of its coal export revenues came from China and India, the world's first and second largest importers of coal, respectively (Chatham House 2021). Nonetheless, the government has sought to lower the country's dependence on coal demand in other countries by creating incentives for domestic industry and power generation to use coal to maximise coal extraction profits (CAT 2020). The country also has significant gas and oil exports totalling US$12.8 billion in 2019 (Chatham House 2021; see also Figure 3.1). However, due to growing domestic demand and falling production, Indonesia has been a net importer of crude oil since 2004 and could become a net importer of gas in the 2020s (Karlis and Rusmana 2017).

Palm oil is the country's second most valuable export commodity, with revenues totalling US$16.9 billion and accounting for 16.7 percent of all foreign exchange earnings

Figure 3.1 Status of fossil fuel and palm oil exports of Indonesia in 2019/2020.

Sources: Data gathered from https://data.worldbank.org (fossil fuel exports as a percentage of merchandise export; oil, gas, and coal as a percentage of GDP – accessed on 25 July 2021) and European Commission 2021b (exports to EU – accessed on 25 July 2021).

from commodity trade in 2019 (Chatham House 2021). The 30 million tons of palm oil Indonesia produces annually generate 4.5 percent of its GDP and provide jobs to 3 million people (UNDP 2018). China, India, and Pakistan are the largest importers of palm oil, but collectively EU member states imported as much as India, the second most important importer, did in 2019, with the import valued at US$2.4 billion (Chatham House 2021; see also Figure 3.1). In Europe, palm oil is mainly used in food, animal feed, and industry, as well as for transportation, with 46 percent of imports being used for biodiesel in 2015 (Copenhagen Economics 2018). As oil palms are higher yielding than other similar crops, they are likely to continue to play an important role in meeting growing global demand for vegetable oils (Meijaard et al. 2018). Timber and other forestry products are also significant exports to Europe, generating US$229 million in 2019 (Chatham House 2021).

Increases in domestic oil demand in the transport sector mean that Indonesia may become more vulnerable to the downsides of fuel subsidies in future. Engaged in oil, gas, and renewable energy production and refining (Natural Resource Governance Institute 2018), state-owned energy company Pertamina bears the costs of subsidising fuel for

Indonesian consumers. Since mismanagement and underinvestment in the oil sector led Indonesia to become a net importer of crude and refined oil in 2004, state revenues have been linked to global oil prices via subsidies on imported fuels, with increases in the global oil price necessitating greater spending on fuel subsidies to keep domestic prices stable (Chelminski 2018). Energy subsidies to reduce and stabilise energy and fuel costs have typically accounted for 10 to 20 percent of government spending and are an important reason why spending on key development areas such as health, education, and public infrastructure has lagged behind other Southeast Asian and lower-middle-income countries (Beaton et al. 2018: 138–139).

On taking office in 2014, the new administration of Joko Widodo was widely praised for reducing fossil fuel subsidies by 90 percent, lowering government spending on fuel subsidies from more than 3 percent of GDP in 2014 to less than 1 percent in 2016 (IEA 2016). However, in 2020 Indonesia still ranked eighth in the world in terms of its spending on energy subsidies, in particular for the oil sector (IEA 2020). New subsidies have also since been introduced. In March 2018, the government capped domestic coal prices at US$70 per tonne, 30 percent below the market rate, until late 2019 – a subsidy that is estimated to cost US$630 million and negatively impact government revenues from coal (Indonesia Investments 2018; Suzuki 2018) – and in April 2018 it introduced controls on fuel and electricity prices. Although artificially keeping fuel prices low has been justified on the basis that it preserves consumer purchasing power, subsidies stimulate overconsumption of imported petrol, which both weakens the rupiah and purchasing power by increasing inflation (Varagur 2018). Even worse, these artificially cheap fossil fuel prices provide an incentive for a less sustainable, more carbon-intensive development path.

Indonesia's rising energy demands from industry and households mean that state-owned power utility Perusahaan Listrik Negara (PLN) and private sector investors are set to make huge investments to expand electricity generation capacity over the coming decade. Renewable energies are supposed to play a stronger role in the overall energy mix. The Medium-Term National Development Plan (RPJMN) 2020–2024 projects reaching 37.3 GW of renewable power capacity in 2024 (CAT 2020), which would mean a more than threefold increase from the roughly 10 GW installed in 2019 (IRENA 2021). This means an increase in ambition compared to early 2019 as PLN's 2019 electricity supply plan had only projected an increase of 8 GW until 2024 (CAT 2020). Given that installed power plant capacity stood at 67.7 GW in 2019 and annual capacity growth from 2018 to 2019 was 4.4 percent (IRENA 2021), these plans are highly ambitious. They could also potentially result in significant idle capacity, given that Indonesia has consistently overbuilt capacity (CAT 2020). However, coal is still accounting for 48 percent of new generation capacity. Both plans envisioned renewables accounting for only around a quarter of new power generation capacity up to 2027 (PWC 2018: 12), despite the falling costs of key renewable energy technologies. In May 2021, the energy ministry announced, however, that it would not approve any new coal-fired power plants in the future, but plants already under construction would be completed (Bloomberg 2021).

The response measures by the Indonesian government to address the Covid-19 pandemic and decreasing energy demand also present a mixed picture in terms of the overall sustainability of its energy sector. The government adopted different measures to subsidise the energy sector. On one hand, coal mining was made eligible to receive fiscal incentives to reduce the impact of Covid-19. On the other, the government provided tax incentives to the solar industry among others (IISD 2020). Nevertheless, the demand for solar PV panels dropped by 70 percent during the pandemic, which is partly explained by overall reduction of government spending in this area (CAT 2020).

In view of the long-term nature of these investments, taking a more carbon-intensive path is likely to present high stranded assets risks and major opportunity costs. The International Renewable Energy Agency (IRENA) already estimated in 2017 that upstream energy production could account for over 80 percent of potentially stranded assets in Indonesia, and in 2015 the Carbon Tracker Initiative calculated that unneeded capital expenditure investments could total almost US$92 billion under a 450-ppm scenario[3] of the International Energy Agency (IEA), with 62 percent of coal investments and 54 percent of investments in natural gas at risk of stranding (Carbon Tracker Initiative 2015; Ivleva et al. 2017: 26). Indonesia's Nationally Determined Contribution (NDC) foresees coal contributing at least 30 percent of the electricity mix in 2025 and 25 percent in 2050 (Government of Indonesia 2016). Furthermore, policies to promote coal are locking in large amounts of investment and high-carbon capacity – with all the associated costs for health and climate change impacts – when these funds could be used to further expand cleaner forms of energy production. Under the contracts drawn up between PLN and independent coal power producers, PLN commits to paying for the contracted amount of electricity whether it is required or not (Chung 2017; Sanchez 2017). The Indonesian state could therefore be locked into funding coal power for decades to come, regardless of future advances and price drops in low-carbon-energy technologies: to be in line with the objectives of the Paris Agreement, Indonesia's coal-fired installed capacity would need to decrease to about 10 percent of its current installed level before 2030 (CAT 2020).

The rapid expansion of coal also has significant health implications, particularly for those living on the islands of Java and Sumatra where most coal-fired power plants are located. The situation may be particularly grave in Jakarta, with 22 coal power plant units operating within 100 kilometres of the city in 2017, and seven more units planned (Greenpeace 2017). Non-communicable diseases caused by air pollution have been found to be a leading cause of premature death in Indonesia (IISD 2018), and it is estimated that every year 100,000 Indonesians die early as a result of poor air quality, caused at least in part by the current energy system (Wright 2016, cited by Ivlelva et al. 2017: 27). When the cost of health and climate change impacts are taken into account, coal power generation may cost up to US$61.5 per MWh – over twice the current market price of coal generation in Indonesia (IISD 2017).

Other fragility and security risks

Indonesia has become increasingly secure and stable over the 2010s. However, growing wealth inequality, ongoing security challenges, and climate change impacts could put pressure on politics and society in future (see Figure 3.2).

Since the Asian financial crisis and the overthrow of H. Muhammed Suharto's 30-year dictatorship in 1998, Indonesia has made steady progress in developing the political institutions of a stable, fully functioning democracy. There is an established pluralism in politics, the media is 'largely free of government control and sanction', and a variety of civil society

Fragility (2021)	Sustainable	Stable	**WARNING**	Alert
Human development (2019)	Very high	**HIGH**	Medium	Low
Strength of governance (2019)	Very high	High	**MEDIUM**	Low
Climate change vulnerability (2019)	Low	Medium	**HIGH**	Very high
Sustainable energy development (2019)	Very high	**HIGH**	Medium	Low

Figure 3.2 Governance indicators (Indonesia).

Sources: Fund for Peace (2021), ND-GAIN (2020), UNDP 2020, World Bank (2021k, 2021l).

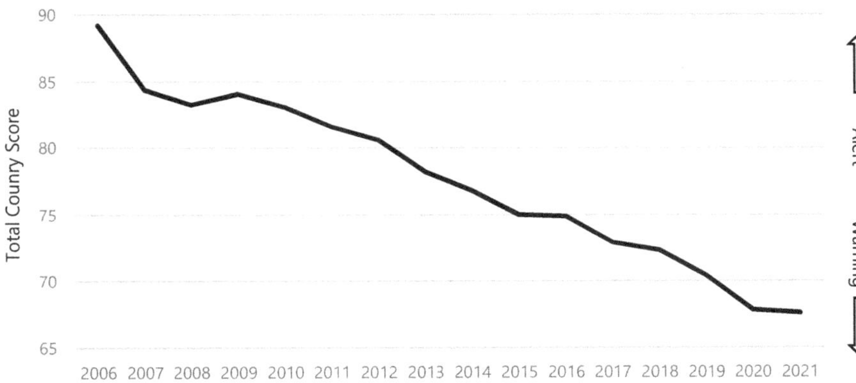

Figure 3.3 Indonesia's performance in the Fragile States Index 2006–2021.

Note: Higher index numbers indicate a higher level of state fragility.

Source: Fund for Peace (2021).

organisations supporting democracy, human rights, and the environment are actively driving this trend (McWilliams 2018). The Freedom House Index of 2021 categorises Indonesia as 'partly free', and Indonesia's position in the Fragile States Index also shows strong improvement in state legitimacy and public services until 2020 (i.e., before the Covid-19 pandemic) (Freedom House 2021; Fund for Peace 2021; see also Figures 3.2 and 3.3).

However, the country also continues to face many of the same challenges it did 20 years ago. An elite class who benefited from association with the Suharto regime remains largely intact, including those who are linked to Indonesia's still-powerful military, which has seen a strengthening of its political role in the 2010s (Bertelsmann Stiftung 2020). Accordingly, the country's growing economic wealth has been concentrated in relatively few hands. Indonesia has one of the highest rates of wealth inequality[4] in the world, and the gap between rich and poor has widened, putting social cohesion under stress. Almost half the population is living in poverty or just above the poverty line, and there are major differences in the provision of basic services like health and education across the country (McWilliams 2018).

Despite continued high approval ratings (Rizki and Cahya 2018), the performance of President Widodo's administration in fulfilling his promises on improving social justice and delivering democratic progress, combating corruption, and addressing human rights issues has remained ambivalent. While many headline development indicators show improvement since he was first elected in 2014, Indonesia's score on the Transparency International Corruption Perceptions Index has improved very little, and the country ranked 102 out of 180 countries in 2020 (Transparency International 2020). According to the Bertelsmann Foundation, 'protection of civil rights remained volatile between 2017 and 2019,' and journalists, government opposition, other social groups, non-mainstream religious groups, and critics of conservative interpretations of Islam have been intimidated and legally prosecuted (Bertelsmann Stiftung 2020). The Fragile States Index also indicates continued human rights issues (Fund for Peace 2021; see also Figure 3.3).

Multiple security threats at the national level pose a risk to democracy and stability in the Indonesian archipelago. At the national level, Indonesia's long history of separatist movements, clashes between divergent ethnic and religious groups, and growing discontent over unequal treatment of indigenous groups have led to tensions in the past. Dissident movements are particularly active in the Aceh and Papua region. Observers also warn about the 'threat of a military coup in the event of a political crisis' as a consequence of persistent scepticism among military leaders with regards to the 'merits of civilian democracy' (Russel 2016).

Despite increased efforts by the Indonesian government to fight terrorism after the 2002 Bali bombing, terrorism remains a palpable threat to security, as the attacks of 2016, 2017, and 2018 have highlighted. In addition, the potential border crossing of extremists and terrorist groups from the Philippines and the rehabilitation of returning Indonesian foreign fighters from Syria represent an important challenge for the Indonesian security landscape (Gupta and Ayesh 2018).

Piracy, illegal fishing, natural disasters, and people and drug-trafficking bear further conflict potential (Sukma 2012). Although the number of piracy incidents has been decreasing since 2015, 26 piracy attacks and attempts were recorded in 2020 (Statista 2021b). Intrusions of the country's maritime borders and hijacking and kidnapping incidents in the surrounding waters have affected Indonesia's maritime security in the past (Gindarsah 2017).

With the second longest coastline in the world, Indonesia is also vulnerable to the effects of climate change and is likely to be severely impacted by changing rainfall patterns, temperature and sea-level rise, and extreme weather events and natural disasters. It has been estimated that climate change impacts in the areas of agriculture, health, and sea-level rise could cost 132 trillion Indonesian rupiahs (equivalent to approximately US$8.9 billion) by 2050, with the capital Jakarta potentially experiencing a quarter of these total projected costs (Hecht 2016).

Concerning agriculture and food security, temperature increase and sea level rise may result in reduced rice productivity and increased damage to crops (US AID 2017), food deficits, and food production loss. As a consequence, the number of undernourished people is likely to increase (Case et al. 2007). Climatic changes will also decrease water quality, increase salinisation of coastal aquifers (US AID 2017), and diminish groundwater resources, drinking water supplies, and irrigation water for agriculture (The Dutch Sustainability Unit 2018).

The increased flood risk due to changing rainfall patterns is likely to increase the pollution of local sources of clean water (Dreierstad 2018). Warming waters, damage to reefs and mangroves, and the erosion of coastlines will likely also contribute to a decreasing marine fish population (US AID 2017), endangering the livelihoods and increasing the vulnerability of coastal communities. It is expected that climate change will also increase risk of forest fires and a loss of habitats. It will thus affect both land and ocean ecosystems (US AID 2017).

Furthermore, the rise in sea level will increase the frequency of floods in Jakarta's and Bekasi's densely populated areas. Runaway development and a near total lack of planning mean that Jakarta is already the world's fastest sinking city, with around 40 percent of the city now below the sea level. Some estimates predict 95 percent of Jakarta's northern port district, home to 1.2 million people, being submerged by 2050 (Kimmelman 2017). As stated in Indonesia's NDC, climate change-induced natural disasters particularly affect people below the poverty line, preventing asset accumulation; provoking rising food, water, and energy prices; and ultimately further reinforcing the already existing poverty. As a consequence, the resulting strong socioeconomic disparity might lead to political instability (Government of Indonesia 2016). Health impacts are also a concern due to the potential for increased incidence of vector- and waterborne diseases, such as dengue fever and malaria (US AID 2017).

Past and present efforts to decarbonise

Indonesia is party to all major climate treaties, having ratified the UN Framework Convention on Climate Change (UNFCCC) in 1994, the Kyoto Protocol in 2004, and the Paris Agreement in 2016. It is a non-Annex I Party to the UNFCCC and is considered a developing country in climate treaties. Indonesia has submitted three National Communications under the UNFCCC, the most recent one in 2018 (as of mid-2021). By 2021, it has also issued two Biennial Update Reports.

In its NDC to the Paris Agreement, the country has committed to a 29-percent emission reduction relative to a business-as-usual scenario by 2030 and, if international support is provided, to increase these efforts to ensure a reduction of 41 percent. Priority areas for reducing emissions outlined in Indonesia's NDC cover both sustainable forest management – a forest moratorium and REDD+ and the transition towards a mixed-energy-use policy with at least 23 percent renewable energy in 2025 – and waste management (Government of Indonesia 2016). Indonesia has not updated or upgraded its original NDC (as of mid-2021).

Indonesia accounts for around 2 percent of global greenhouse gas (GHG) emissions (WRI 2021). To date, energy and land use, land use change, and forestry (LULUCF) have been by far the highest emitting sectors (WRI 2021; see Figure 3.4). Having lost a total of 15 percent of its forest cover due to land use change and peat fires between 2001 and 2017 (Global Forest Watch 2018), reducing deforestation and implementing afforestation programmes will be crucial for both adaptation and mitigation. Palm oil is the largest single cause of deforestation, and palm oil plantations are also reducing biodiversity, having been found to have harmed over 190 threatened species (Carrington 2018). However, the country's rapid industrial and urban development means that future emissions growth will in large part

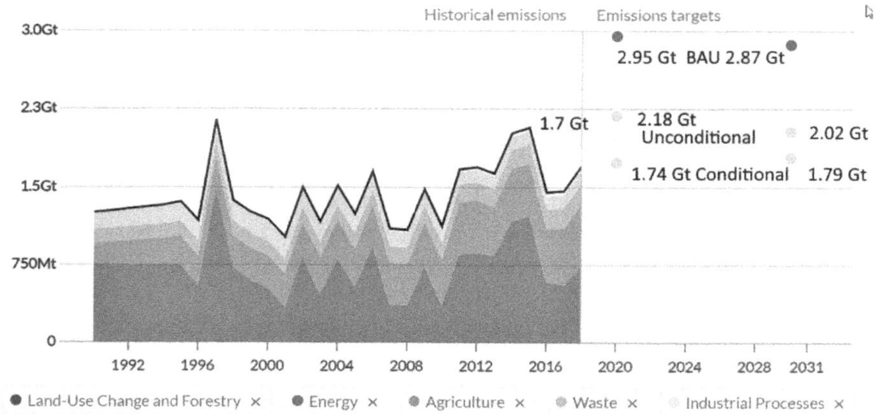

Figure 3.4 Indonesia's greenhouse gas emissions and emission targets (1990–2018).
Source: WRI (2021).

come from industry, transportation, and electricity supply (Rissmann 2017). Analysis by the World Resources Institute forecasts that energy emissions will outstrip LULUCF emissions and account for over half of total emissions by 2026 (Wijaya et al. 2017).

Indonesia's long-term development priorities, as expressed in its national development plan for 2005 to 2025, are squarely focused on stimulating economic growth to lift more people out of poverty and on strengthening its democracy and international standing. Climate and environmental protection, and their contribution to long-term prosperity, are not explicitly mentioned among these core objectives.

Climate policy falls under the remit of the Ministry of Forestry and Environment. Several policies detailing Indonesia's climate action efforts exist, including the National Medium Term Development Plans for 2015–2019 and for 2020–2024 (RPJMN), with efforts to foster a green economy and promote sustainable infrastructures. As part of the new 2020–2024 Plan, the government has launched a Low-Carbon Development Initiative to ensure the environmental credentials of the next five-year plan (GCEC 2018), and a low-carbon development plan, including three emission pathways to 2045 (CAT 2020). The National Action Plan for Greenhouse Gas Emission Reduction and the National Adaptation Action Plan on Climate Change are the country's core mitigation and adaptation policies. They present Indonesia's strategic vision and specify areas for action on green cities, rural development, natural resources, community resilience, forestry and peat lands, agriculture, energy and transportation, industry, and waste (GRICCE 2017).

Current policy projections for sectors beyond LULUCF show that the country will easily overachieve its current NDC targets, and this likely overachievement has led the ambition of its NDC target (excluding forestry) to be rated as 'highly insufficient' (CAT 2020). Indeed, a large proportion of emission reductions up to 2030 set to come from the forestry sector and Indonesia's forest moratorium is the policy with the greatest potential to reduce emissions. It has been calculated that if it is renewed in its current form up to

2030, it could lower emissions by around 199 Mt CO_2. Furthermore, if the moratorium were extended to cover more forest and forest areas under concession licenses, emissions might be reduced by as much as 437 Mt CO_2 by 2030 – the equivalent to shutting off 108 coal power plants for a year (WRI 2020; calculated using EPA 2017). This would require better enforcement of existing policies and regulations, with more resources for combatting illegal deforestation and mining (UNDP 2018).

To date, there has been strong government, private sector, and public support for Indonesia's current development path, which exploits its abundant domestic coal reserves and promises low-cost energy in the short term. A 2018 survey of stakeholders in the Indonesian power sector found that they prioritised the affordability of energy and security of supply over sustainability (PWC 2018: 30). This focus on keeping the price of energy low reflects mainstream public opinion. The huge increases in coal-fired power generation, including in and around major cities, has not prompted any significant backlash in Indonesia – unlike in other nearby countries like Thailand where local population has protested and staged hunger strikes against new coal power plants (Palma 2018) – and a 2015 survey found that 79 percent of Indonesians were opposed to energy subsidy reform, despite its benefits (Pfefferle 2018). Although sustainable development was an important manifesto issue for Jokowi in the presidential elections in April 2019 (Heriyanto 2019), the affordability of energy was a much more prominent issue in both candidates' campaigns. The government's cap on domestic coal prices and price controls on fuel and electricity were widely considered to be aimed at boosting Widodo's popularity, and other candidates were reluctant to announce measures that would raise the cost of electricity or fuel (Crooks 2018). However, in the medium term, as the price of renewable energy technologies continues to fall, there is likely to be less of a trade-off between affordability and sustainability (PWC 2018).

However, there are powerful players supporting the continued development of fossil fuels in Indonesia. Its history of oil production and fossil fuel subsidies has given rise to an '"oil and gas mafia", known for its corruption, including embezzlement of funds from the Ministry of Energy, extortion, tax fraud and smuggling' (Cassin 2014; Sukoyo 2014), and the coal industry has a strong lobby in Indonesian politics and business (Crooks 2018). 'State-owned oil company Pertamina, the Indonesian oil-trading lobby, vehicle manufacturers and distributors and freight and public transport' and other vested interests in industries that benefit most from subsidised fuels have lobbied intensively against them. Pertamina, for example, would be unable to compete in the market for refined oil products without subsidies due to decades of insufficient investment in the company (Chelminski 2018).

Trends and potential

While the country's growth model has brought prosperity to a rising number of Indonesians, Indonesia's challenge is to leapfrog the carbon-intensive development pattern taken by the West and China and to move towards sectors with a more sustainable, long-term future. How well and how quickly Indonesia achieves this could also have implications more globally as other emerging economies and developing countries in Southeast Asia and Africa look to Indonesia as a model for successful economic development (Rissmann 2017).

Financing low-carbon development

Putting Indonesia on a sound footing in a decarbonised world will require sustained political will and increased investments in renewable energy technologies and efficient equipment. As shown above, Indonesia is already set to invest enormous sums into its energy system over the coming decades. Greater ambition in investing in clean energy technologies could significantly reduce stranded asset risks, and, even before the costs of health impacts and climate change are factored in, these investments would more than pay for themselves over time (Rissmann 2017).

As well as reducing vulnerability to shocks on international markets and increasing spending on public goods such as healthcare and education, further reducing subsidies on fuel and on electricity generated by fossil fuels could support Indonesia's low-carbon transition in at least two important ways. First, reducing government spending on these subsidies could enhance scope for increasing investment in energy efficiency, renewable energy technologies, and other low-carbon sectors of the economy. Second, reducing subsidies can improve efficiency, preventing unnecessary investments in expensive new power stations. The most recent National Medium Term Development Plan for 2020–2024 (RPJMN) has taken one step in this direction, but it remains to be seen to what extent the support for traditional energy sector players as a response to the pandemic crisis will counteract this approach.

Interim results from decreasing subsidies have indicated some positive trends. The increases in electricity prices resulting from lower power-sector subsidies have put electricity consumption on a lower-growth trajectory (Varagur 2018). Between its 2017 and 2018 business plan, state utility PLN reduced its target for new power generation capacity down from 78 GW to 56 GW by 2027, not least due to lower expected average energy demand growth[5] (PWC 2018). Fuel subsidies have increased demand for fuel and contributed to the rapid spread of privately owned motor vehicles (Asian Development Bank 2015; IEA 2016). The transport sector is responsible for almost 40 percent of Indonesian energy demand, and this is also where the greatest future increase is to be expected. Consequently, reducing subsidies has been shown to significantly reduce traffic in Indonesia's congested cities (Burke et al. 2017).

Subsidy reform has had a chequered history in Indonesia – the steep reductions in subsidies and fuel price hikes following the Asian financial crisis in 1998 resulted in riots that contributed to the toppling of the Suharto regime. However, successful reforms, such as those enacted by Jokowi in 2014, have shown that to ensure social acceptance, reforms must be accompanied by social assistance to ensure that poorer households are not disadvantaged by rising energy prices. Strong political leadership and strategically communicating reform to the public have also been important (Chelminski 2018). Further reductions of subsidies initiated before the Covid-19 pandemic hold the potential to realise some of the potential indicated above (IISD 2021).

Potential in key low-carbon sectors

Indonesia accounts for over 36 percent of energy demand in Southeast Asia, making it the largest energy consumer in the region (IEA 2021a). With energy demand rising and electricity demand potentially doubling until 2040 (IEA 2021a), efficiency measures are essential to keep energy use and expenditure down and reduce emissions. If current policies are implemented and enforced, they could result in energy savings of 2 percent per year up to 2025 (IEA 2017). However, there is ample scope for increasing ambition, with Indonesia scoring only 35 out of 100 for energy efficiency in the World Bank's Regulatory Indicators for Sustainable Energy Index (World Bank 2020b). Important areas for action include cooling and transport. Average yearly installations of air-conditioning units are increasing at a rate of 13 percent in Indonesia, making the country the second fastest growing market behind India (Cool Coalition 2020). The total number of air-conditioning units in Southeast Asia is expected to rise from 40 million units in 2017 to about 300 million units in 2040 and 50 percent of the new units are expected to be installed in Indonesia (IEA 2019). If implementation of regional targets for energy efficiency in the cooling sector were to be accelerated, Indonesia could avoid an additional 32 petajoules (PJ) of electricity consumption by 2030, equivalent to 5.2 million barrels of oil. With regard to the transport sector, if Indonesia matched China in terms of fuel efficiency standards for heavy duty vehicles, and if electric motorcycles were to achieve the same market penetration as in China, Indonesia could reduce its fuel imports by 13 percent by 2030 and reduce its spending on oil imports by over US$1.4 billion per year (IEA 2017; see also IEA 2021a). Regulatory frameworks and a lack of infrastructure are also holding back progress in electric mobility (IRENA 2017).

Regarding renewable energy, Indonesia is a country with 'phenomenal resources for renewable power generation' (IRENA 2017). Aside from emission reductions, renewable energy poses major advantages for improving air quality in cities, reducing health hazards, and supplying electricity to Indonesia's more remote provinces and islands. There are also important financial co-benefits – taking into account the impacts of climate change and health costs of air pollution – scaling up renewables can save the country between US$15.6 billion and US$51.7 billion per year (IRENA 2017). Some technologies are also increasingly able to compete with domestic coal on costs. In 2019, installed capacity from renewable sources totalled 9,861 MW, accounting for 14.6 percent of the country's electricity capacity and generation (IRENA 2021, PWC 2018). In its business plan for developing Indonesia's electricity supply for the period 2018–2027, state-owned electricity utility PLN envisages adding almost 15 GW of new renewable generation capacity by 2027, which would represent 27 percent of new capacity overall (PWC 2018). As mentioned above, the National Development Plan 2020–2024 foresees an upgrade of this expansion target (adding more than 25 GW new renewable generation capacity by 2024). Indonesia's NDC states that renewable energy will provide at least 23 percent of the country's primary energy supply by 2025 and at least 31 percent by 2050.

In 2019 hydropower and geothermal power accounted for almost 80 percent of electricity from renewable sources. Hydropower provides the greatest proportion, with 5,616 MW of installed capacity covering over 6 percent of national electricity production in 2019 (IRENA 2021). The business plan foresees around 8 GW of new power generation capacity being added by 2027, although the Ministry of Energy and Mineral Resources estimates that potential hydropower generation could be as high as 75 GW[6] (PWC 2018). Indonesia also has the world's second largest reserves of geothermal energy, with potential power generation of 29 GW (PWC 2017). In 2019, just above 2 GW of power – around 3 percent of the Indonesian energy mix – was generated by geothermal plants (IRENA

2021), with 4.5 GW of new power generation capacity set to be added by 2027 (PWC 2017, 2018). The development of this form of energy is being driven by state-owned enterprises (PWC 2017). Both these technologies provide base-load renewable power that can balance a high share of intermittent renewable energy sources, such as wind and solar PV. However, in 2017 Indonesia still trailed other G20 countries in installed capacity of wind and solar PV (Allianz 2017). Whereas solar and wind respectively contributed a mere 198 MW and 76 MW of generation capacity in 2019 (IRENA 2021), ministry assessments found installed peak capacity could total 207.8 GWp from solar PV and 60.6 GW from wind power (Government of Indonesia 2019). In the latest business plan, PLN proposed raising its target for renewables other than hydropower and geothermal energy to 2,046 MW, which would represent only 4 percent of the projected new generation capacity by 2027 (PWC 2018).

To accelerate growth in renewables, both government and private sector investment will be needed. However, the power sector is still heavily regulated by the government, and there have been numerous changes to regulations for independent power producers, creating uncertainty and holding back investment (IRENA 2017; Allianz 2017). As such, Indonesia ranks 19 on the Allianz Climate and Energy Investment Index for G20 countries, which evaluates policy adequacy, policy reliability, market absorption capacity, and general involvement conditions[7] (Allianz 2018). Private sector investors are demanding high rates of return before entering or increasing their stake in the Indonesian renewables market (Allianz 2017; PWC 2018). Further barriers include the fragmented nature of the grid across the Indonesian archipelago, which makes effective policy and planning difficult, and the bottlenecks in PLN's transmission and distribution systems, which is unlikely to be able to accommodate high amounts of distributed solar power without significant investment (IRENA 2017). Ensuring the continued and successful rollout will therefore require more detailed long-term policy and investment planning. Indonesia has been one of four countries in the G20 without a long-term plan to decarbonise the grid and power system (Allianz 2018).

Education and skills development

Indonesia has made huge progress raising the level of education among its young and growing population, particularly in the past 15 years. However, according to the World Bank's Human Capital Index, the current level of education and healthcare would allow the average child born in Indonesia today to achieve around 54 percent of its potential, below the average for the East Asia and Pacific region and below the average for upper-middle-income countries (World Bank 2020a). Indonesia still lags behind neighbouring countries like Vietnam and Singapore in the average years of schooling and the quality of its education system (World Bank 2018). While it ranked 36 out of 137 participating countries in the 2017–2018 Global Competitiveness Index overall, it only came in the 94th place for primary education and health and 64th place for higher education and training (WEF 2017). The literacy rate in Indonesia is 95.7 percent, and the average Indonesian attends school for 14 years (CIA 2021). Enrolment in both secondary education and tertiary education has steadily increased to 89 percent and 36 percent, respectively, in 2018 (World Bank 2021h, 2021i).

Services was the fastest growing sector in Indonesia in the period from 2007 to 2017, achieving an average annual growth of 7 percent (ILO 2017; Statista 2021a), and employed around 50 percent of the population in 2019. Agriculture remains a significant employer, but its share of the workforce has gradually fallen from 39 percent in 2010 to 28 percent in 2020 (Statista 2021a), despite the sector growing on average 4 percent per year (ILO 2017). Just over a fifth of Indonesians are employed in industry, which includes mining, manufacturing, and construction (ILO 2017; Statista 2021a). Although the general pre-pandemic unemployment rate was relatively low (3.6 percent in 2019) (World Bank 2021d), youth unemployment stood at 20 percent (World Bank 2021e). Due to the pandemic, unemployment rose to 6.3 percent as of February 2021 (*The Jakarta Post* 2021). The economy is also facing further challenges regarding the quality of work, gender, and disparities across the provinces (ILO 2017).

Technological advancements are shaping the economy in Indonesia, and digitisation will likely lead to both the loss and creation of jobs. The e-commerce, on-demand services, and transportation sectors are growing and offering new employment opportunities. However, the 'technology versus jobs debate' continues, and further education and training schemes will become increasingly important as increasing numbers of jobs are automated, especially since many Indonesians only have low-level educational qualifications (ILO 2017).

Cooperation with the EU

As the largest economy in Southeast Asia and the seat of ASEAN, Indonesia is an increasingly important partner for the EU, and a sound basis has already been developed for cooperation between Indonesia and the EU. Trade in fossil fuels between Indonesia and the EU is relatively insignificant. By contrast, the EU is a major importer of Indonesian palm oil, which is used in many sectors, including for biofuels. This has given rise to tensions given EU intentions to phase out the use of palm oil in the production of biofuels. Sustainable forest management has been a focal area for development cooperation.

Institutional framework for cooperation with Indonesia and ASEAN

EU-Indonesia relations have developed rapidly in recent years. The Partnership and Cooperation Agreement that entered into force in 2014 established a firm framework for cooperation in the fields of trade, tourism, energy, transport, environment, and natural resources, as well as in security and combating terrorism (European Commission 2009; EU 2014). In addition, the EU and Indonesia hold regular political dialogues, human rights dialogues, and security dialogues (EU-Indonesia 2018). For example, the EU and Indonesia are working towards advancing the EU-Indonesia Security and Defence Partnership (EEAS 2018a).

Negotiations for a Comprehensive Economic Partnership Agreement (CEPA) began in 2016 with the aim of achieving an 'ambitious and mutually beneficial trade agreement' (EEAS 2018b; see also EEAS 2018c). The agreement is set to cover a broad range of issues, including tariffs and non-tariff barriers to trade, as well as trade in services and investment (European Commission 2021a). It is also set to include a sustainability chapter. As of mid-2021, negotiations are still ongoing.

The EU has also deepened its ties with ASEAN. Already, in 2015, the High Representative of the Union for Foreign Affairs and Security Policy issued a Joint Communication entitled 'The EU and ASEAN: a partnership with a strategic purpose' (European Commission 2015) which specifies climate change action as a key focal area for strengthening interregional cooperation. Initiatives are underway to support sustainable forest and peatland management in ASEAN countries. These include measures to the Sustainable Use of Peatlands and Haze Mitigation in ASEAN (SUPA) programme, and the EU FLEGT (Forest Law Enforcement, Governance and Trade) Facility to promote the trade of legally produced timber. In 2013, the latter resulted in a Voluntary Partnership Agreement between the EU and Indonesia to assure access of legal timber exports to the EU market (European Commission 2016). In December 2020, the EU and ASEAN elevated their partnership with a commitment to regular summits at leaders' level (ASEAN/EU 2020).

Ultimately, the EU hopes to use its bilateral trade negotiations and agreements with ASEAN countries as steppingstones towards an EU-ASEAN trade agreement. Free trade agreements have already been concluded with Singapore (2014) and Vietnam (2015), and, alongside Indonesia, negotiations are ongoing with Malaysia, Thailand, and the Philippines (European Commission 2021a). In 2018, the High Representative also issued a Joint Communication proposing several policy initiatives and proposals to serve as building blocks for an EU strategy for improving connectivity between the EU and Asia, including improving transport, energy, and digital links (European Commission 2018a). An example of an EU-ASEAN initiative that will feed into this strategy is the €85 million the EU has provided to support for consolidating the ASEAN community via a single market and reduced non-tariff barriers to trade (European Commission 2018b).

Trade

In 2020, the bilateral trade in goods between the EU and Indonesia amounted to €20.6 billion, down from €23.8 billion in 2019. In services, bilateral trade amounted to €7.5 billion in 2019 (European Commission 2021a). The EU mainly exports high-tech machinery, transport equipment, manufacturing goods, and chemicals to Indonesia (EEAS 2017; see also European Commission 2021b). For example, half of Indonesia's defence imports throughout the period 2004–2017 were from the EU and Switzerland (ISS 2018). All in all, the EU-27 accounts for 5.5 percent of Indonesian commodities exports, with the Netherlands (1 percent) as the biggest export partner (Chatham House 2021). Trade in fossil fuels between Indonesia and the EU is relatively insignificant, with EU coal and oil exports accounting for less than 0.5 percent of Indonesian imports and Indonesian coal exports to the EU having decreased from 3.2 percent in 2011 to less than 1 percent of EU coal imports in 2019 (Chatham House 2021).

Far more significant are Indonesia's crude palm oil exports to the EU, which is its second largest export market (see Figure 3.1). The EU has been 'the most open market for Indonesian palm oil' (Copenhagen Economics 2018), with Indonesia providing 36 percent of all EU palm oil imports in 2019 (Chatham House 2021). Almost half of EU palm oil imports are used for the production of biofuels (Copenhagen Economics 2018). The issue of palm oil in European biofuels has become a focal issue as the EU 2018 Renewable Energy Directive foresees the gradual phase-out of food crops such as palm oil that it alleges result in high indirect land use change (EEAS 2018d; Voegele 2018). This promises to severely restrict Indonesian palm oil exports to the EU and has been an issue in bilateral relations, including in the CEPA negotiations. As a sign of the growing tensions, Indonesia launched a WTO lawsuit on the matter in late 2019 (European Commission 2019; Nangoy and Munthe 2019). In order to settle the dispute, a panel was established on the request of Indonesia in 2020 (see https://www.wto.org/english/tratop_e/dispu_e/cases_e/ds593_e.htm).

Development cooperation

The EU and its member states have worked in partnership with Indonesia on a range of initiatives to achieve the Sustainable Development Goals (SDGs) and tackle climate change (EEAS 2018e). Due to Indonesia achieving G20 member status, the EU is no longer providing Indonesia with bilateral aid. However, the EU has supported Indonesia's national climate change response via a programme, which has particularly supported the Aceh and Papua provinces in their efforts towards low-emission development planning and enhancing education (EEAS 2016). The EU also provided emergency humanitarian assistance after the Sulawesi earthquake and tsunami in September 2018. The total collective EU aid amounted to €18 million. In addition, the EU deployed civil protection and humanitarian experts to help coordinate relief efforts on site (EEAS 2018f). This is complemented by bilateral aid by EU member states as well as in the broader ASEAN context.

Conclusions

With its abundant natural resources and as one of the world's most populous countries, Indonesia has advanced to become a regional leader and a major economy. Since the 1980s, it has diversified its economy beyond fossil fuels, although carbon intensive exports, such as coal and palm oil, remain important pillars of its economic development. Furthermore, as a fast-developing nation with a growing population and rising expectations in terms of income and living standards, electricity and fuel consumption is increasing rapidly, and their affordability is an important political issue. The Indonesian government and state utility have invested heavily in exploiting the country's abundant hydrocarbons, particularly coal, to meet these energy needs. While this may be considered cost-effective at present, there are major health and opportunity costs to following the high-carbon development path taken by the West and China, rather than leapfrogging more quickly to low-carbon development models. As Indonesia is a role model for other developing countries in Southeast Asia and Africa, the way it develops and how quickly it is able to harness its high energy efficiency and renewable energy potential to move to a low-emission economy may also have wider implications for global emission reductions.

Efforts to green the European transport sector and reduce the impact of biofuels on deforestation and emissions in other parts of the world may present a risk for Indonesian palm oil exports. Continued exports to Europe are likely to rely on Indonesia achieving a balance between developing its economically important palm oil industry and measures to protect its tropical forests and reduce its high emissions from land use and land use change. This also applies to Indonesian timber exports to Europe. With regard to the power sector, coal consumption and power sector decarbonisation trends in India and China will have an important influence on Indonesian coal export revenues. Oil imports and the pressure of continued – albeit reduced – fuel subsidies on government budgets leave Indonesia exposed to fluctuations in the international oil price, which is increasingly being influenced by decarbonisation trends. In the past, oil price increases led to domestic price shocks and political unrest in Indonesia.

The EU and Indonesia are cooperating to improve the sustainability of Indonesia's palm oil and timber exports, for example, via strengthened diplomacy, certification schemes, and a strong focus on sustainable forest management in development cooperation. Furthermore, in the negotiations for a CEPA, the EU and Indonesia aim to establish a broader basis for future EU-Indonesia relations and a deeper economic partnership. This will support the development of other areas of mutual interest for future EU-Indonesia relations, including diversified trade and sustainable development. The EU and its member states can assist in important ways to help Indonesia accelerate its transition to a low-carbon economy.

On the basis of these findings, we can identify a number of prominent focal points for the EU to support Indonesia's low-carbon development under European and global decarbonisation:

- **EU-Indonesia Comprehensive Economic Partnership Agreement**: The conclusion of the CEPA with Indonesia can provide a basis for increased and more diversified trade with Indonesia. Maintaining an open and constructive dialogue with the Indonesian government and stakeholders throughout the negotiations, and particularly on sensitive issues such as definitions of sustainability and the role of palm oil and forestry products, will be key to the success of the final agreement.
- **Sustainable forest management**: The EU can continue to provide assistance and work with Indonesian stakeholders to strengthen the Voluntary Partnership Agreement on timber exports by improving the system assuring the legality of timber, its implementation, and monitoring of the agreement's social, economic, and environmental effects.
- **Renewable energy**: The EU and its member states have substantial expertise in fields such as renewable energy and energy efficiency in industrial processes and households. They can therefore provide positive support for Indonesia's low-carbon energy transition in the form of capacity-building and technology transfer. Offering economically viable alternative pathways is essential, as domestic availability of coal – and in the forest sector the economic importance of palm oil production – are seemingly first-best solutions for Indonesia to develop its economy. Finance and technology transfer can improve the attractiveness of a lower-carbon pathway.
- **Promote just transition narratives in EU (climate) diplomacy**: The EU and its member states have developed policies and good practices to support coal-producing regions diversify and create low-carbon jobs, as well as gathering lessons learned. The EU can encourage exchange with Indonesian counterparts on this issue. The EU and its member states can also seek closer cooperation with Indonesia within international and multilateral forums promoting the low-carbon energy transition, including in the context of the recovery from the Covid-19 pandemic.

- **Promote sustainable urban development**: European cities have also tested a range of policies to reduce traffic congestion and air pollution in urban areas. European diplomacy can encourage exchange on this issue as European best practices (and failures) may be a valuable source of knowledge for Indonesian cities as they grow and the country becomes increasingly urbanised.

Overall, Indonesia is an increasingly important partner for the EU, as it is both a major economy and a leading voice within ASEAN. The EU has established a fruitful basis for cooperation, and the ongoing CEPA negotiations are creating the potential for further cooperation and trade in existing and new areas, including many that are relevant for decarbonisation. Building a consensus on the sensitive issue of palm oil sustainability in biofuels will be important to the success of the agreement and to future relations, given its significance for Indonesian trade with the EU – and for shaping decarbonisation pathways on both sides. Indonesia and the EU also share common challenges with regard to decarbonisation, such as sustainable urban development and the low-carbon energy transition. This includes ensuring the just transition of regions dependent on the production of coal and other high-carbon products, and the rollout of high proportions of renewable energy technologies. These areas can form the basis of deepened cooperation as the EU and Indonesia pursue more ambitious decarbonisation policies.

Notes

1 GDP per capita increased from US$2,144 in 2000 to US$4,451 in 2019 (in constant 2010 US$; World Bank 2021b).
2 Measured in terms of the poverty headcount ratio at US$1.90 a day (2011 PPP) (World Bank 2021c).
3 Introduced in the International Energy Agency (IEA) 2014 World Energy Outlook (IEA 2014), the 450 ppm scenario – compatible with limiting warming to 2°C – assumes that oil demand will peak around 2020.
4 'Indonesia's wealth Gini coefficient – 84 on a 0-to-100 scale – is among the highest in the world and has gone up since 2012. Income disparity is comparatively almost as severe (62nd) and has deepened since 2012' (WEF 2018).
5 From 8.3 percent in the 2017 business plan to 6.9 percent in the 2018 version (PWC 2018: 12).
6 Although the Ministry's technical assessments of potential power generation do not take into account financial/economic viability (PWC 2018).
7 In the Allianz Climate and Energy Investment Index for the G20, Indonesia ranks fourth highest in terms of investment needs and 14th in terms of investment attractiveness, for which it scores 31 out of 100 (Allianz 2017: 40).

References

Allianz (2017) *Allianz Climate and Energy Monitor 2017. Assessing the Needs and Attractiveness of Low-Carbon Investments in G20 Countries*. Available at: https://newclimate.org/wp-content/uploads/2017/07/allianz-climate-and-energy-monitor-2017-report_final.pdf (accessed 25 September 2018).

Allianz (2018) *Allianz Climate and Energy Monitor 2018. Assessing the Needs and Attractiveness of Low-Carbon Investments in G20 Countries*. Available at: https://www.allianz.com/content/dam/onemarketing/azcom/Allianz_com/sustainability/documents/Allianz_Climate_and_Energy_Monitor_2018.pdf (accessed 9 January 2019).

Ariffin, E. (2018) 'The Decline of Indonesia's Oil and Gas Industry', *The Asean Post*, 13 May 2018. Retrieved 25 September 2018 from https://theaseanpost.com/article/decline-indonesias-oil-and-gas-industry.

ASEAN/EU (2020) '"Co-chairs" Press Release of the 23rd ASEAN-EU Ministerial Meeting'. Retrieved 26 July 2021 from https://www.consilium.europa.eu/en/press/press-releases/2020/12/01/co-chairs-press-release-of-the-23rd-asean-eu-ministerial-meeting/.

Asian Development Bank (2015) *Fossil Fuel Subsidies in Indonesia. Trends, Impacts, and Reforms*. Available at: from https://www.adb.org/sites/default/files/publication/175444/fossil-fuel-subsidies-indonesia.pdf (accessed 2 October 2018).

Beaton, C., Lontoh, L., and Wai-Poi, M. (2018) 'Indonesia: Pricing reforms, social assistance, and the importance of perceptions'. In: Inchauste, G. and Victor, D.G. (eds.) (2017) *The political economy of energy subsidy reform*. Washington D.C.: World Bank.

Bertelsmann Stiftung (2020) *BTI 2020. Indonesia Country Report 2020*. Available at: https://bti-project.org/content/en/downloads/reports/country_report_2020_IDN.pdf (accessed 13 July 2021).

Bloomberg (2021) 'No New Coal Plants in Indonesia in Another Bid to Cut Emissions', 27 May 2021. Retrieved 26 July 2021 from https://www.bloomberg.com/news/articles/2021-05-27/no-new-coal-plants-in-indonesia-in-another-bid-to-cut-emissions.

BP (2020) *BP Statistical Review of World Energy 2020*. Available at: https://www.bp.com/content/dam/bp/business-sites/en/global/corporate/pdfs/energy-economics/statistical-review/bp-stats-review-2021-full-report.pdf (accessed 1 July 2021).

Burke, J.P., Batsuuri, T., and Halley Yudhistira, M. (2017) 'Easing the traffic: The effects of Indonesia's fuel subsidy reforms on toll-road travel', *Transportation Research Part A: Policy and Practice* 105(C), pp. 167–180. https://doi.org/10.1016/j.tra.2017.08.003.

Carbon Tracker Initiative (2015) *The $2 Trillion Stranded Assets Danger Zone: How Fossil Fuel Firms Risk Destroying Investor Returns*. Available at: http://www.carbontracker.org/wp-content/uploads/2015/11/CAR3817_Synthesis_Report_24.11.15_WEB2.pdf (accessed 14 August 2017).

Carrington, D. (2018) 'Palm oil "disastrous" for wildlife but here to stay, experts warn', *The Guardian*, 26 June 2018. Retrieved 25 September 2018 from https://www.theguardian.com/environment/2018/jun/26/palm-oil-disastrous-for-wildlife-but-here-to-stay-experts-warn.

Case, M., Ardiansyah, F., and Spector, E. (2007) 'Climate Change in Indonesia. Implications for Humans and Nature'. Prepared on behalf of WWF. Retrieved 25 September 2018 from http://assets.wwf.org.uk/downloads/indonesian_climate_ch.pdf.

Cassin, R.L. (2014) 'Indonesia Anti-Graft Agency to Take on "Oil and Gas Mafia"'. Retrieved 26 September 2018 from https://fcpablog.com/2014/12/2/indonesia-anti-graft-agency-to-take-on-oil-and-gas-mafia/.

CAT – Climate Action Tracker (2020) 'Indonesia'. Retrieved 13 July 2021 from https://climateactiontracker.org/countries/indonesia/current-policy-projections/.

Chatham House (2021) 'Resource Trade. Earth'. Retrieved 13 July 2021 from http://resourcetrade.earth/.

Chelminski, K. (2018) 'Fossil fuel subsidy reform in Indonesia'. In: Van Asselt, H. and Skovgaard, J. (eds.) *The politics of fossil fuel subsidies and their reform*. Cambridge: Cambridge University Press, pp. 193–211.

Chung, Y. (2017) *Overpaid and Underutilized: How Capacity Payments to Coal-Fired Power Plants Could Lock Indonesia into a High-Cost Electricity Future*. Prepared on behalf of IEEFA. Available at: from http://ieefa.org/wp-content/uploads/2017/08/Overpaid-and-Underutilized_How-Capacity-Payments-to-Coal-Fired-Power-Plants-Could-Lock-Indonesia-into-a-High-Cost-Electricity-Future-_August2017.pdf (accessed 25 September 2018).

CIA (2021) 'The World Factbook. Indonesia'. Retrieved 13 July 2021 from https://www.cia.gov/the-world-factbook/countries/indonesia/#introduction.

Cool Coalition (2020) 'IEA: Additional Efforts on Cooling Needed to Reduce Buildings-Related Emissions'. Retrieved 14 July 2021 from https://coolcoalition.org/iea-additional-efforts-on-cooling-needed-to-reduce-buildings-related-emissions/.

Copenhagen Economics (2018) *EU Imports of Palm Oil from Indonesia, Malaysia and Thailand*. Malaysian Palm Oil Council. Available at: https://www.copenhageneconomics.com/dyn/resources/Publication/publicationPDF/8/448/1528720336/eu-imports-of-palm-oil-16may2018.pdf (accessed 10 December 2018).

Crooks, E. (2018) 'Coal fades in developed world but is far from dead in Asia', *Financial Times*, 14 June 2018. Retrieved 25 September 2018 from https://www.ft.com/content/cf717854-6818-11e8-aee1-39f3459514fd.

Deloitte (2017) 'Taxation and Investment in Indonesia 2017'. Available at: https://www2.deloitte.com/content/dam/Deloitte/global/Documents/Tax/dttl-tax-indonesiaguide-2017.pdf (accessed 25 September 2018).

Dreierstad, I. (2018) *How Climate Change Affects Food and Water Security in Indonesia*. Future Directions International. Available at: https://www.futuredirections.org.au/wp-content/uploads/2018/02/How-Climate-Change-Affects-Food-and-Water-Security-in-Indonesia.pdf (accessed 25 September 2018).

European Commission (2009) 'Framework Agreement on Comprehensive Partnership and Cooperation between the European Community and its Member States, of the One Part, and the Republic of Indonesia, of the Other Part'. Available at: http://collections.internetmemory.org/haeu/20160313172652/http:/eeas.europa.eu/delegations/indonesia/documents/eu_indonesia/eu_idnpca_en.pdf (accessed 25 September 2018).

European Commission (2015) *Joint Communication to the European Parliament and the Council. The EU and ASEAN: A Partnership with a Strategic Purpose*. JOIN(2015) 22 final. Available at: https://eur-lex.europa.eu/legal-content/EN/TXT/PDF/?uri=JOIN:2015:22:FIN&from=EN (accessed 12 November 2018).

European Commission (2016) *Environment: EU and Indonesia to Start First-Ever Licensing Scheme for Legal Timber Exports*. Available at: http://ec.europa.eu/environment/pdf/21_04_2016_en.pdf (accessed 25 September 2018).

European Commission (2018a) 'Connecting Europe and Asia - Building blocks for an EU Strategy. Joint Communication to the European Parliament, the Council, the European Economic and Social Committee, the Committee of the Regions and the European Investment Bank'. JOIN(2018) 31 final. Available at: https://eeas.europa.eu/sites/eeas/files/joint_communication_-_connecting_europe_and_asia_-_building_blocks_for_an_eu_strategy_2018-09-19.pdf (accessed 7 January 2019).

European Commission (2018b) *Connecting Europe & Asia. The EU Strategy*. Available at: https://cdn2-eeas.fpfis.tech.ec.europa.eu/cdn/farfuture/Nqd1l36P_4zLcndw5kEdAOq7lG0kGs_degOz-7TprJc/mtime:1537345558/sites/eeas/files/europe_asia_connectivity_factsheet_1.pdf (accessed 7 January 2019).

European Commission (2019) *Report of the 8th Round of Negotiations for a Free Trade Agreement between the European Union and Indonesia*. Available at: http://trade.ec.europa.eu/doclib/docs/2019/july/tradoc_158207.pdf (accessed 29 July 2019).

European Commission (2021a) 'Countries and Regions. Indonesia'. Retrieved 14 July 2021 from https://ec.europa.eu/trade/policy/countries-and-regions/countries/indonesia/.

European Commission (2021b) 'European Union. Trade in Goods with Indonesia'. Retrieved 14 July 2021 from https://webgate.ec.europa.eu/isdb_results/factsheets/country/details_indonesia_en.pdf.

EEAS – European External Action Service (2016) 'Indonesia and the EU'. Retrieved 25 September 2018 from https://eeas.europa.eu/generic-warning-system-taxonomy/404/680/indonesia-and-eu_en.

EEAS (2017) 'EU-Indonesia Relations'. Retrieved 25 September 2018 from https://eeas.europa.eu/headquarters/headquarters-homepage/4009/eu-indonesia-relations_fi.

EEAS (2018a) 'Speech by H.E. Vincent Guérend, the EU Ambassador to Indonesia, at the Public Dialogue "Advancing EU-Indonesia Security and Defence Partnership"'. Retrieved 17 December 2018 from https://eeas.europa.eu/delegations/indonesia/54676.

EEAS (2018b) 'Report from the 4th Round of Negotiations for a Comprehensive Economic Partnership Agreement (CEPA) between the EU and Indonesia'. Retrieved 25 September 2018 from https://eeas.europa.eu/headquarters/headquarters-homepage/41460/report-4th-round-negotiations-comprehensive-economic-partnership-agreement-cepa-between-eu-and_de.

EEAS (2018c) 'Stakeholders Share Views on Expected Impacts of the EU-Indonesia CEPA'. Retrieved 25 September 2018 from https://eeas.europa.eu/headquarters/headquarters-homepage/49411/stakeholders-share-views-expected-impacts-eu-indonesia-cepa_de.

EEAS (2018d) 'Palm Oil: Outcome of the Trilogue of the EU's Renewable Energy Directive (RED II)'. Retrieved 25 September 2018 from https://eeas.europa.eu/delegations/indonesia/46646/palm-oil-outcome-trilogue-eu%E2%80%99s-renewable-energy-directive-red-ii_en.

EEAS (2018e) *Blue Book 2018: EU-Indonesia*. From Development Cooperation. Available at: https://reliefweb.int/sites/reliefweb.int/files/resources/eu-indonesia_blue_book_2018.pdf (accessed 18 July 2021).

EEAS (2018f) 'Sulawesi Earthquake & Tsunami: EU Solidarity in Action'. Retrieved 17 December 18 from https://eeas.europa.eu/headquarters/headquarters-homepage/51699/sulawesi-earthquake-tsunami-eu-solidarity-action_ja.

Elias, S. and Noone, C. (2011) *The Growth and Development of the Indonesian Economy*. Available at: https://www.rba.gov.au/publications/bulletin/2011/dec/pdf/bu-1211-4.pdf (accessed 25 September 2018).

EPA – United States Environmental Protection Agency (2017) 'Greenhouse Gas Equivalencies Calculator'. Retrieved 25 September 2018 from https://www.epa.gov/energy/greenhouse-gas-equivalencies-calculator.

EU – European Union (2014) 'Framework Agreement on Comprehensive Partnership and Cooperation between the European Community and Its Member States, of the One Part, and the Republic of Indonesia'. Available at: http://trade.ec.europa.eu/doclib/docs/2016/july/tradoc_154810.pdf (accessed 25 September 2018).

EU – Indonesia (2018) *Blue Book 2018*. Available at: https://eeas.europa.eu/sites/eeas/files/eu-idn_blue_book_2018.pdf (accessed 25 September 2018).

Freedom House (2021) 'Freedom in the World 2021'. Indonesia. Retrieved 13 July 2021 from https://freedomhouse.org/country/indonesia/freedom-world/2021.

Fund for Peace (2021) 'Fragile States Index 2021'. Retrieved 26 July 2021 from https://fragilestatesindex.org/.

Gindarsah, I. (2017) 'Indonesia's Security and Defense Outlook: Evolving Low-Intensity Threats and Recalibrated Countermeasures'. National Institute for Defense Studies. Retrieved 26 September 2018 from http://www.nids.mod.go.jp/english/publication/joint_research/series15/pdf/chapter02.pdf.

GCEC – Global Commission on the Economy and Climate (2018) 'Press Release: Partnership Towards Indonesia's Science-Based Low Carbon Development'. Retrieved 24 October 2018 from https://newclimateeconomy.net/content/press-release-partnership-towards-indonesias-science-based-low-carbon-development.

Global Forest Watch (2018) 'Country Profile Indonesia'. Retrieved 30 July 2018 from https://www.globalforestwatch.org/dashboards/country/IDN?category=summary.

Government of Indonesia (2016) *First Nationally Determined Contribution Republic of Indonesia*. Available at: 30 July 2018 from http://www4.unfccc.int/ndcregistry/PublishedDocuments/Indonesia%20First/First%20NDC%20Indonesia_submitted%20to%20UNFCCC%20Set_November%20%202016.pdf (accessed 30 July 2018).

Government of Indonesia (2019) *Indonesia energy outlook 2019*. Jakarta: MEMR.

GRICCE – Grantham Research Institute on Climate Change and the Environment (2017) 'Country Profile: Indonesia'. Retrieved 30 July 2018 from http://www.lse.ac.uk/GranthamInstitute/country-profiles/indonesia/.

Greenpeace (2017) *Jakarta's Silent Killer*. Available at: https://www.greenpeace.org/static/planet4-southeastasia-stateless/2020/10/48d34259-jakartas-silent-killer-report-oct-24-2017-1.pdf (accessed 25 September 2018).

Gupta, A. and Ayesh, O. (2018) 'A guide to navigating the 2018 risk landscape in Indonesia', *The Diplomat*, 29 January 2018. Retrieved 26 September 2018 from https://thediplomat.com/2018/01/a-guide-to-navigating-the-2018-risk-landscape-in-indonesia/.

Hecht, J.E. (2016) *Indonesia: Costs of Climate Change 2050*. USAID. Available at: https://www.climatelinks.org/sites/default/files/asset/document/Indonesia%20Costs%20of%20CC%202050%20Policy%20Brief.pdf (accessed 25 September 2018).

Heriyanto, D. (2018) 'Here are Jokowi-Ma'ruf's nine "missions" for 2019's presidential poll', *The Jakarta Post*, 5 December 2018. Retrieved 26 July 2019 from https://www.thejakartapost.com/news/2018/12/04/here-are-jokowi-marufs-nine-missions-for-2019s-presidential-poll.html.

IEA – International Energy Agency (2014) 'World Energy Outlook'. Retrieved 24 September 2018 from https://www.iea.org/publications/freepublications/publication/WEO2014.pdf.

IEA (2016) 'Indonesia's Steady Progress in Tackling Fossil Fuel Subsidies'. Retrieved 25 September 2018 from https://www.iea.org/newsroom/news/2016/december/indonesias-steady-progress-in-tackling-fossil-fuel-subsidies.html.

IEA (2017) *Energy Efficiency 2017*. Available at: https://www.iea.org/publications/freepublications/publication/Energy_Efficiency_2017.pdf (accessed 25 September 2018).

IEA (2019) 'The Future of Cooling in Southeast Asia. Increasing Energy Efficiency through Stronger Policy Action'. Retrieved 14 July 2021 from https://www.iea.org/reports/the-future-of-cooling-in-southeast-asia.

IEA (2020) 'Energy Subsidies. Tracking the Impact of Fossil-Fuel Subsidies'. Retrieved 13 July 2021 from https://www.iea.org/topics/energy-subsidies.

IEA (2021a) 'E4 Country Profile: Energy Efficiency Indonesia'. Retrieved 14 July 2021 from https://www.iea.org/articles/e4-country-profile-energy-efficiency-indonesia.

IEA (2021b) 'Indonesia. Data Browser. Energy Consumption. Electricity Consumption'. Retrieved 14 July 2021 from https://www.iea.org/countries/indonesia.

IEA (2021c) 'Indonesia. Data Browser. Energy Supply. Electricity Generation by Source'. Retrieved 13 July 2021 from https://www.iea.org/countries/indonesia.

IISD – International Institute for Sustainable Development (2017) *Financial Supports for Coal and Renewables in Indonesia GSI Report*. Available at: https://www.iisd.org/sites/default/files/publications/financial-supports-coal-renewables-indonesia.pdf (accessed 25 September 2018).

IISD (2018) *The Health Cost of Coal in Indonesia GSI Report*. Available at: https://www.iisd.org/sites/default/files/publications/health-cost-coal-indonesia.pdf (accessed 25 September 2018).

IISD (2020) *Indonesia's Energy Policy Briefing*. Available at: https://www.iisd.org/system/files/2020-08/indonesia-energy-policy-briefing-july-2020-en.pdf (accessed 27 July 2021).

IISD (2021) *Liquefied Petroleum Gas Subsidy Reform in Indonesia: Lessons Learned from International Experience*. Available at: https://www.iisd.org/system/files/2021-05/lpg-subsidy-reform-indonesia.pdf (accessed 26 July 2021).

ILO – International Labour Organization (2017) *Indonesia Jobs Outlook 2017. Harnessing Technology for Growth and Job Creation*. Available at: https://www.ilo.org/wcmsp5/groups/public/---asia/---ro-bangkok/---ilo-jakarta/documents/publication/wcms_613628.pdf (accessed 25 September 2018).

Indonesia Investments (2018) 'Domestic market obligation Indonesia: Coal price capped at $70 per ton', *Indonesia Investments*, 11 March 2018. Retrieved 25 September 2018 from https://www.indonesia-investments.com/business/business-columns/domestic-market-obligation-indonesia-coal-price-capped-at-70-per-ton/item8653?.

IRENA (2017) *Renewable Energy Prospects: Indonesia. Executive Summary*. Available at: http://www.irena.org/-/media/Files/IRENA/Agency/Publication/2017/Mar/IRENA_REmap_Indonesia_summary_2017.pdf?la=en&hash=F530E18BAFC979C8F1A0254AFA77C9EBC9A0EC44 (accessed 25 September 2018).

IRENA (2021) 'Energy Profile Indonesia'. Retrieved 13 July 2021 from https://www.irena.org/IRENADocuments/Statistical_Profiles/Asia/Indonesia_Asia_RE_SP.pdf.

ISS – European Union Institute of Security Studies (2018) *Guns, Engines and Turbines. The EU's Hard Power in Asia*. Available at: https://www.iss.europa.eu/sites/default/files/EUISSFiles/CP_149_Asia.pdf (accessed 21 December 2018).

Ivleva, D., Schlösser, T., Scholl, C., Schultze, K., and Wolters, S. (2017) *Alte Schätze nicht mehr heben? Stranded Assets und ihre Bedeutung für die Governance des fossilen Rohstoffsektors*. Bonn/Eschborn: Deutsche Gesellschaft für Internationale Zusammenarbeit (GIZ) GmbH. Available at: https://www.adelphi.de/de/system/files/mediathek/bilder/Alte%20Schätze%20nicht%20mehr%20heben%20Stranded%20Assets%20und%20ihre%20Bedeutung%20für%20die%20Governance%20des%20fossilen%20Rohstoffsektors%20-%20GIZ%20adelphi.pdf (accessed 18 July 2021).

Karlis, S. and Rusmana, Y. (2017) 'Whatever happened to Indonesia's mighty oil and gas industry?' *Bloomberg*, 14 August 2017. Retrieved 25 September 2018 from https://www.bloomberg.com/news/articles/2017-08-14/whatever-happened-to-indonesia-s-mighty-oil-and-gas-industry.

Kimmelman, M. (2017) 'Jakarta is sinking so fast, it could end up underwater', *The New York Times*, 21 December 2017. Retrieved 19 October 2018 from https://www.nytimes.com/interactive/2017/12/21/world/asia/jakarta-sinking-climate.html.

McWilliams, E. (2018) 'Democracy in Indonesia: A progress report', *The Foreign Service Journal*, May 2018. Retrieved 25 September 2018 from http://www.afsa.org/democracy-indonesia-progress-report.

Meijaard, E., Garcia-Ulloa, J., Sheil, D., Wich, S.A., Carlson, K.M., Juffe-Bignoli, D., and Brooks, T.M. (2018) *Oil Palm and Biodiversity. A Situation Analysis by the IUCN Oil Palm Task Force*. Prepared on behalf of IUCN. Switzerland: IUCN. Available at: https://portals.iucn.org/library/sites/library/files/documents/2018-027-En.pdf (accessed 18 July 2021).

Nangoy, F. and Munthe, B.C. (2019) 'Indonesia defends palm oil after EU targets 2030 phase-out in road fuel', *Reuters*, 13 February 2019. Retrieved 31 July 2019 from https://in.reuters.com/article/us-indonesia-palmoil-eu-idINKCN1Q21P3.

Natural Resource Governance Institute (2018) 'Indonesia. Oil and Gas'. Retrieved 25 September 2018 from https://www.resourcegovernanceindex.org/country-profiles/IDN/oil-gas.

ND-GAIN – Notre Dame Global Adaptation Initiative (2020) 'ND-GAIN Country Index'. South Bend, IN, USA: University of Notre Dame. Retrieved 5 July 2021 from https://gain.nd.edu/our-work/country-index/.

Palma, S. (2018) 'Indonesia keeps faith with coal to meet growing energy demand', *Financial Times*, 31 July 2018. Retrieved 25 September 2018 from https://www.ft.com/content/0fd3ed3e-89b9-11e8-affd-da9960227309.

Pfefferle, T. (2018) 'Fossil fuel subsidy reform in Indonesia: A complex case', *Medium*, 15 November 2017. Retrieved 25 September 2018 from https://medium.com/@tim.pfefferle/fossil-fuel-subsidy-reform-in-indonesia-a-complex-case-e99b23503f47.

PWC – PricewaterhouseCoopers and The Lagos Chamber of Commerce and Industry (2017) *Power in Indonesia. Investment and Taxation Guide*. Available at: https://www.pwc.com/id/en/energy-utilities-mining/assets/power/power-guide-2017.pdf (accessed 25 September 2018).

PWC (2018) *Alternating Currents: Indonesian Power Industry Survey 2018*. Available at: https://www.pwc.com/id/en/publications/assets/eumpublications/utilities/power-survey-2018.pdf (accessed 25 September 2018).

Rissmann, J. (2017) 'How to decarbonize Indonesia and achieve low-carbon prosperity', *Forbes*, 15 August 2017. Retrieved 26 September 2018 from https://www.forbes.com/sites/energyinnovation/2017/08/15/how-to-decarbonize-indonesia-and-achieve-low-carbon-prosperity/#66fbb27c4f1f.

Rizki, N. and Cahya, Y. (2018) 'Unofficial counts show regional Indonesia polls favoring President Widodo', *Reuters*, 27 June 2018, Retrieved 26 September 2018 from https://www.reuters.com/article/us-indonesia-elections/unofficial-counts-show-regional-indonesia-polls-favoring-president-widodo-idUSKBN1JN024.

Russel, M. (2016) *At a Glance. Indonesia: Security Threats to a Stable Democracy*. Prepared on behalf of the European Union. European Union. Available at: https://www.europarl.europa.eu/RegData/etudes/ATAG/2016/593481/EPRS_ATA(2016)593481_EN.pdf (accessed 18 July 2021).

Sanchez, L. (2017) 'How Indonesia Might Turn Its Back on a Future of Cheaper Renewable Electricity'. Global Subsidies Initiative. Retrieved 26 September 2018 from https://www.iisd.org/gsi/subsidy-watch-blog/how-indonesia-might-turn-its-back-future-cheaper-renewable-electricity.

Singgih, V.P. (2017) 'Indonesia happy to stay outside of OPEC for now', *The Jakarta Post*, 6 December 2017. Retrieved 26 September 2018 from http://www.thejakartapost.com/news/2017/12/06/indonesia-happy-to-stay-outside-of-opec-for-now.html.

Statista (2021a) 'Indonesia: Distribution of Employment by Economic Sector from 2010 to 2020'. Retrieved 14 July 2021 from https://www.statista.com/statistics/320160/employment-by-economic-sector-in-indonesia/.

Statista (2021b) 'Number of Actual and Attempted Piracy Attacks in Indonesia from 2008 to 2020'. Retrieved 23 July 2021 from https://www.statista.com/statistics/250866/number-of-actual-and-attempted-piracy-attacks-in-indonesia/.

Sukma, R. (2012) 'Indonesia's Security Outlook, Defence Policy and Regional Cooperation'. Retrieved 26 September 2018 from http://www.nids.mod.go.jp/english/publication/joint_research/series5/pdf/5-1.pdf.

Sukoyo, Y. (2014) 'For Jokowi, oil mafia is a slick target to catch', *Jakarta Globe*, 23 September 2018. Retrieved 26 September 2018 from https://jakartaglobe.id/news/jokowi-oil-mafia-slick-target-catch/.

Suzuki, W. (2018) 'Indonesia to lose $630m in revenue from coal price cap', *Nikkei Asia*, 13 March 2018. Retrieved 26 September 2018 from https://asia.nikkei.com/Economy/Indonesia-to-lose-630m-in-revenue-from-coal-price-cap.

The Dutch Sustainability Unit of the Netherlands Commission for Environmental Assessment (2018) *Climate Change Profile: Indonesia*. Available at: https://reliefweb.int/sites/reliefweb.int/files/resources/Indonesia_2.pdf (accessed 15 July 2021).

The Jakarta Post (2021) 'Unemployment remains high at 6.26% as millions continue to reel from pandemic "fallout"', 7 May 2021. Retrieved 14 July 2021 from https://www.thejakartapost.com/news/2021/05/07/unemployment-remains-high-at-6-26-as-millions-continue-to-reel-from-pandemic-fallout.html.

Transparency International (2020) 'Corruption Perception Index 2020'. Retrieved 13 July 2021 from https://fragilestatesindex.org/country-data/.

UNCTAD – United Nations Conference on Trade and Development (2019) 'State of Commodity Dependence 2019'. Retrieved 13 July 2021 from https://unctad.org/fr/node/27405.

UN DESA – United Nations Department of Economic and Social Affairs Population Division (2018) 'World Urbanization Prospects 2018'. Retrieved 13 July 2021 from https://population.un.org/wup/DataQuery/.

UN DESA (2019) 'World Population Prospects 2019'. Retrieved 13 July 2021 from https://population.un.org/wpp/DataQuery/.

UNDP – United Nations Development Programme (2018) 'Indonesia Palm Oil'. Retrieved 22 July 2021 from https://www1.undp.org/content/dam/gp-commodities/docs/Country%20Factsheets/INDONESIA%20PALM%20OIL.pdf?download.

UNDP (2020) Human Development Report. Available at: http://hdr.undp.org/en/countries (accessed 26 July 2021).

US AID (2017) 'Climate Risk Profile Indonesia'. Retrieved 26 September 2018 from https://www.climatelinks.org/sites/default/files/asset/document/2017_USAID_ATLAS_Climate%20Risk%20Profile_Indonesia.pdf.

Varagur, K. (2018) 'Indonesia to effectively continue fuel subsidy', *Voanews*, 19 March 2018. Retrieved 26 September 2018 from https://www.voanews.com/a/indonesia-fuel-subsidy/4304709.html.

Voegele, E. (2018) 'EU reaches deal on REDII, sets new goals for renewables', *Biodiesel Magazine*, 15 June 2018. Retrieved 26 September 2018 from http://www.biodieselmagazine.com/articles/2516392/eu-reaches-deal-on-redii-sets-new-goals-for-renewables.

WEF – World Economic Forum (2017) *The Global Competitiveness Report 2017–2018*. Available at: http://www3.weforum.org/docs/GCR2017-2018/05FullReport/TheGlobalCompetitivenessReport2017%E2%80%932018.pdf (accessed 14 July 2021).

WEF (2018) 'Selected Country Summaries'. Retrieved 26 September 2018 from http://reports.weforum.org/the-inclusive-development-index-2018/selected-country-summaries/.

Wijaya, A., Chrysolite, H., Ge, M., Wibowo, C.K., Pradana, A., Firselly Utami, H., and Austin, K. (2017) 'How Can Indonesia Achieve Its Climate Change Mitigation Goal? An Analysis of Potential Emissions Reductions from Energy and Land-Use Policies'. Working Paper prepared on behalf of the World Resources Institute. Washington D.C.: World Resources Institute. Available at: https://files.wri.org/d8/s3fs-public/how-can-indonesia-achieve-its-climate-change-mitigation-goal-analysis-potential-emissions-reductions-from-energy-land-use-policies_0.pdf (accessed 26 September 2018).

World Bank (2018) *Indonesia Economic Quarterly. Learning Faster, Growing Faster*. Indonesia Economic Quarterly (IEQ). Available at: http://documents.worldbank.org/curated/en/305361528210283009/pdf/126891-WP-PUBLIC-on-6-5-18.pdf (accessed 1 August 2018).

World Bank (2020a) 'Human Capital Project October 2020. Indonesia. Human Capital Index 2020'. Retrieved 14 July 2021 from https://databank.worldbank.org/data/download/hci/HCI_2pager_IDN.pdf?cid=GGH_e_hcpexternal_en_ext.

World Bank (2020b) 'RISE. Indonesia'. Retrieved 14 July 2021 from http://rise.worldbank.org/country/indonesia.

World Bank (2021a) 'GDP Ranking. Gross Domestic Product 2020'. Retrieved 13 July 2021 from https://datacatalog.worldbank.org/dataset/gdp-ranking.

World Bank (2021b) 'GDP Per Capita (Constant 2010 US$)'. Retrieved 13 July 2021 from https://data.worldbank.org/indicator/NY.GDP.PCAP.KD?locations=ID.

World Bank (2021c) 'Poverty. Poverty Headcount Ratio at $1.90 a Day (2011 PPP) (% of Population)'. Retrieved 13 July 2021 from https://data.worldbank.org/indicator/SI.POV.DDAY?display=&locations=ID.

World Bank (2021d) 'Unemployment, Total (% of Total Labor Force) (Modeled ILO Estimate)'. Retrieved 13 July 2021 from https://data.worldbank.org/indicator/SL.UEM.TOTL.ZS?locations=ID.

World Bank (2021e) 'Share of Youth Not in Education, Employment or Training, Total (% of Youth Population)'. Retrieved 13 July 2021 from https://data.worldbank.org/indicator/SL.UEM.NEET.ZS?locations=ID.

World Bank (2021f) 'Exports of Goods and Services (% of GDP)'. Retrieved 13 July 2021 from https://data.worldbank.org/indicator/NE.EXP.GNFS.ZS?locations=ID.

World Bank (2021g) 'Fuel Exports (% of Merchandise Exports) – Indonesia'. Retrieved 13 July 2021 from https://data.worldbank.org/indicator/TX.VAL.FUEL.ZS.UN?locations=ID.

World Bank (2021h) 'School Enrollment, Secondary (% Gross) – Indonesia'. Retrieved 14 July 2021 from https://data.worldbank.org/indicator/SE.SEC.ENRR?locations=ID.

World Bank (2021i) 'School Enrollment, Tertiary (% Gross) – Indonesia'. Retrieved 14 July 2021 from https://data.worldbank.org/indicator/SE.TER.ENRR?end=2018&locations=ID&start=1970&view=chart.

World Bank (2021j) 'GDP (Constant 2010 US$) – Indonesia'. Retrieved 14 July 2021 from https://data.worldbank.org/indicator/NY.GDP.MKTP.KD?locations=ID.

World Bank (2021k) 'Worldwide Governance Indicators 2019'. Retrieved 26 July 2021 from http://info.worldbank.org/governance/wgi/#reports.

World Bank (2021l) *Regulatory Indicators for Sustainable Energy (RISE)*. Retrieved 26 July 2021 from https://rise.esmap.org.

WRI – World Resources Institute (2020) 'This Interactive Chart Shows Changes in the World's Top 10 Emitters. Indonesia'. Retrieved 13 July 2021 from https://www.wri.org/insights/interactive-chart-shows-changes-worlds-top-10-emitters.

WRI (2021) 'Climate Watch. Indonesia. Greenhouse Gas Emissions and Emissions Targets'. Retrieved 13 July 2021 from https://www.climatewatchdata.org/countries/IDN#ndc-content-overview.

4 Azerbaijan

Partnership potential beyond gas?

Introduction

The Republic of Azerbaijan (henceforth Azerbaijan) is a country of around 10 million inhabitants (CIA 2021) located in the South Caucasus region between Eastern Europe and Western Asia. Azerbaijan shares borders with Iran, Armenia, Georgia, and Russia, and it is a coastal state to the west of the Caspian Sea. A quarter of the size of Germany, Azerbaijan is composed of 66 administrative divisions with Baku its biggest city and capital. The country is semi-urbanised with 56.8 percent of the population living in towns and cities (CIA 2021). Azerbaijan is a presidential republic headed since 2003 by Ilham Aliyev, in his fourth term, having succeeded his father Haider Aliyev. The country gained independence from the Soviet Union on 30 August 1991 and became part of the Commonwealth of Independent States (CIA 2021). Azerbaijan-Russia ties remain close and friendly.

Azerbaijan's economy has seen dynamic development since the mid-1990s and in particular after 2004, driven by the development of its oil and gas industry. From 1995 to 2019, Azerbaijan's gross domestic product (GDP) increased more than 15-fold to about US$48 billion (in current US$) (World Bank 2021c). GDP per capita similarly grew more than tenfold to about US$4.800 in 2019 (World Bank 2021d). As a result, the percentage of people living under the poverty line decreased from 24 percent in 2005 to around 5 percent from 2015 onwards (Indexmundi 2020; CIA 2021). Unemployment is relatively low at around 5 percent (2019) (Trading Economics 2021a).

Accordingly, Azerbaijan has become an upper-middle-income country, and available indices indicate a medium ranking of Azerbaijan regarding economic welfare and human development. Hence, on the Human Development Index of 2019 Azerbaijan ranks 88 out of 189 countries (UNDP 2020). On the World Economic Forum's Inclusive Development Index of 2018 Azerbaijan ranks third among the emerging and developing economies (World Economic Forum 2018). On the 2021 Sustainable Development Goals Index Azerbaijan is ranked 55 out of 165 countries, progressing at an average pace for its region (Bertelsmann Stiftung 2021). Also, Azerbaijan ranks 71 out of 179 countries on the Fragile States Index in 2021 (Fund for Peace 2021), with a significant improvement especially in the economic dimension since 2009 (see Figure 4.3). Similarly, Azerbaijan overall ranked 98 out of 137 countries on the Bertelsmann Stiftung's Transformation Index, taking 76th place on economic transformation (Bertelsmann Stiftung 2020).

DOI: 10.4324/9781003183037-4

Following the general approach of the case studies, this chapter will explore the challenges and opportunities Azerbaijan and Azerbaijan-EU relations face under decarbonisation in five steps. The next section first demonstrates Azerbaijan's high dependence on the production and export of oil and gas and hence its vulnerability to price fluctuations and the phase-out of fossil fuels. It also explores other fragility and security risks, including the longstanding conflict between Azerbaijan and Armenia over Nagorno-Karabakh. Subsequently, we assess the status of Azerbaijan's climate policy, which leaves much room for further improvement. This is followed by the review of relevant trends and potentials, especially regarding efforts to diversify the economy that have only borne limited fruit so far. The one but last section reviews the framework and basis of EU-Azerbaijan relations beyond fossil fuels. The final section concludes and identifies focal points for EU foreign policy towards Azerbaijan under decarbonisation.

Exposure and risks

Exposure to global decarbonisation trends

Azerbaijan's economy is highly dependent on the production and export of oil and gas. The country depends on oil and gas for at least half of its economy and government revenue. Accordingly, while the development of its oil and gas resources has in general helped to advance welfare and provide for political stability in the twenty-first century, Azerbaijan has remained exposed to oil price fluctuations.

Azerbaijan is a significant producer of oil and gas. In 2019, it produced about 780,000 barrels of oil per day, equivalent to somewhat less than 1 percent of the world total. Oil production has decreased by more than 20 percent since the beginning of the 2010s. Azerbaijan's gas production reached 24.3 billion cubic metres in 2019 (an increase of about 50 percent over the course of the decade), which put Azerbaijan among the top 30 gas-producing countries (BP 2020; CIA 2021; SOCAR 2021a).

Azeri gas reserves are more significant than its oil reserves. Azerbaijan's proven oil reserves are estimated to amount to about 7 billion barrels, which might – given current production levels – be exhausted within the next two to three decades (BP 2020; see also Shepard 2016). Accordingly, Azeri oil production is expected to decline further in the medium term (Gurbanov et al. 2017). In contrast, estimated gas reserves of up to 2.8 trillion cubic metres could support current production levels for more than 100 years (BP 2020; see for other estimates: CIA 2021; EIA 2021).

In accordance with its resource base, Azerbaijan mainly uses gas and oil as a source of energy supply. Overall, domestic oil consumption has been increasing slightly in the 2010s to around 108,000 barrels per day in 2019 (around 14 percent of production) (BP 2020). Similarly, gas consumption has been increasing over the past decade, accounting for about 50 percent of overall production in 2019 (11.8 billion cubic metres: BP 2020), which has also been increasing (see above). Gas (13 power stations) generally provides for more than 90 percent of power production (with some fluctuation depending on the availability of hydropower depending on rainfall). Overall, about two-thirds of overall energy consumption is accounted for by gas, about 30 percent by oil, with the remaining 2–3 percent made up of renewable energy, primarily hydropower (EIA 2019; BP 2020; World Bank 2021a).

Oil and gas exports are of crucial importance for Azerbaijan's economy. In the 2010s, the country exported about 85 percent of its oil production and an increasing share of up to 50 percent of its gas production (BP 2020; CIA 2021). Oil and gas account for over 90 percent of Azeri exports (World Bank 2021g), which makes the country 'vulnerable to external shocks' (European Parliament 2017: 22; see also Figure 4.1).[1]
As a result, oil and gas continue to play a dominant role in the Azeri economy and for the government budget. As the engine of the economy, the sector accounts for about half of GDP (fluctuating somewhat with the oil price) and the state budget. Both shares have tended to decline in the 2010s because of the declining oil production and oil price (see below) and the Azeri government's efforts to diversify the economy, including through the State Oil Fund of the Republic of Azerbaijan (SOFAZ), as well as its attempts to broaden the budget income base. The SOFAZ was created in 1999 to accumulate oil revenues for the benefit of current and future generations. One of its major objectives is to provide support for the development of non-oil sectors (Breban and Mukhtarov 2017; Gurbanov et al. 2017: 6). Overall, the diversification of the economy remains a major challenge nevertheless (see below).

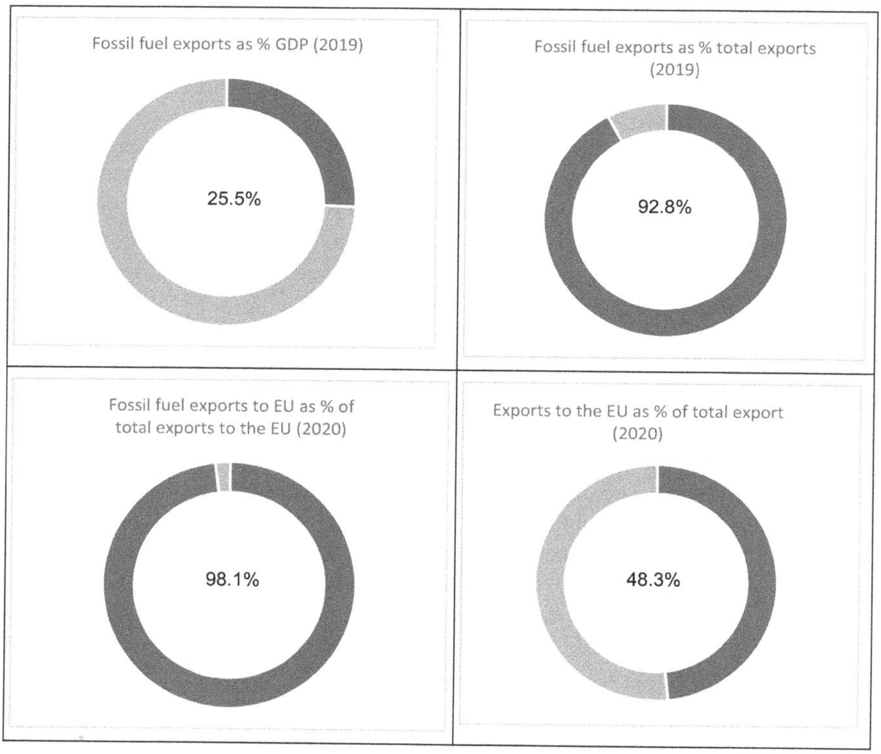

Figure 4.1 Status of fossil fuel exports of Azerbaijan in 2019/2020.
Sources: Data based on: https://data.worldbank.org (fossil fuel exports as a percentage of merchandise export; oil, gas, and coal as a percentage of GDP – accessed on 25 July 2021) and European Commission 2021 (exports to EU – accessed on 25 July 2021).

While relying on the oil and gas sector for its income, the Azeri government also provides subsidies for energy. According to the International Energy Agency, in 2020 US$487 million of subsidies were given to the oil sector, US$236 million for electricity, and US$302 million for natural gas, primarily for consumption (overall amounting to about 2.4 percent of GDP) (IEA 2020).

Despite some tools to manage oil price fluctuations, Azerbaijan has remained exposed to variations in oil (and related gas) prices. Declining oil prices after 2014 have left a clear mark on economic development. GDP in current US$ declined sharply from 75.2 billion in 2014 to 37.9 billion in 2016. It recovered to about US$48 billion in 2019 before receding to US$42.6 billion in 2020 in the wake of the Covid-19 pandemic. GDP per capita similarly dropped from US$7,900 in 2014 to less than US$3,900 in 2016, before increasing to US$4,700 in 2019 (2020: 4200). Due to exchange rate fluctuations, the figures look far less dramatic in constant 2010 US$ according to which overall GDP dropped from US$57.9 billion in 2014 to US$56.7 billion in 2016, reaching US$59.1 billion in 2019 and US$56.5 billion in 2020 (indicating a decline of GDP of nearly 5 percent in 2020). GDP per capita in constant US$ declined from US$6,073 in 2014 to US$5,765 in 2017, reaching US$5,895 in 2019 and US$5,593 in 2020 (World Bank 2021c, 2021d, 2021e, 2021f).

Accordingly, the contribution of revenue from the oil and gas sector to the government budget has also declined. While the SOFAZ has been used as a buffer for balancing the state budget in times of low or falling oil prices (under the banner of 'supporting macroeconomic stability') (Breban and Mukhtarov 2017; Gurbanov et al. 2017: 6), Azerbaijan's public debt has nevertheless risen in the wake of sharp falls in the oil price from 2014 to 2016. The debt-to-GDP ratio rose from 7.3 percent in 2008 to more than 22 percent in 2016, after which it only slightly declined to about 18 percent in 2019 (with the partial recovery of oil prices), before increasing to more than 20 percent in 2020 (as a result of the Covid-19 pandemic) (Trading Economics 2021b). Overall, the debt-to-GDP ratio has hence remained modest (with some exposure to fluctuations).

Azerbaijan's energy infrastructure has remained closely tied to its oil and gas resource base, carrying the danger of them turning into 'stranded assets'. Three oil pipelines serve exports to the Russian and Georgian Black Sea coasts as well as the Turkish Mediterranean harbour of Ceyhan. The Baku-Tbilisi-Ceyhan Pipeline is by far the most important of these pipelines and started operating in the 2000s. It has sufficient capacity (1,200,000 barrels per day) to support current/declining Azeri oil exports. By far the most important gas pipeline is the South Caucasus Pipeline, which runs along the BTC oil pipeline via Georgia to Turkey and started operating in 2007. It has been further developed as the 'Southern Gas Corridor' to connect to Greece and Italy (via Turkey). It thereby directly connects to the EU gas grid and supplies increased Azeri gas production to Europe. Also, gas power stations have been modernised over the past decade (including a further switch from oil to gas) (EIA 2019). The Southern Gas Corridor became operational in 2020.

Enormous investments continue to be made in the oil and gas sector, partially by Azerbaijan itself and partially by foreign private investors (Foreign Direct Investment, FDI). The investments required for the Southern Gas Corridor have been estimated to amount to about €40 billion. FDI in Azerbaijan amounted to nearly US$15 billion in 2017, the lion's share of which went to oil and gas (Santander 2018).

Other fragility and security risks

Contrasting its economic advances, Azerbaijan continues to face relatively serious challenges with respect to corruption, democracy, and the protection of political rights. Accordingly, the country ranks 129 out of 180 countries on the Transparency International Corruption Perceptions Index (Transparency International 2021). The Freedom House Index of 2021 categorises Azerbaijan as 'not free', pointing to serious limitations regarding political rights, civil liberties, and press freedom (Freedom House 2021). Similarly, sub-indicators of the Fragile States Index relating to state legitimacy and human rights remain relatively low (Fund for Peace 2021), which is consistent with the findings of the Bertelsmann Stiftung's Transformation Index of 2020 (Bertelsmann Stiftung 2020; see also Figures 4.2 and 4.3).

Accordingly, the European Union criticised the presidential elections of April 2018 in which President Aliyev won a fourth term of office (and which the opposition boycotted). The EU Spokesperson for Foreign Affairs and Security Policy, including European Neighbourhood Policy and Enlargement Negotiations, stated that the elections

Fragility (2021)	Sustainable	Stable	**WARNING**	Alert
Human development (2019)	Very high	**HIGH**	Medium	Low
Strength of governance (2019)	Very high	High	**MEDIUM**	Low
Climate change vulnerability (2019)	Low	**MEDIUM**	High	Very high
Sustainable energy development (2019)	Very high	High	**MEDIUM**	Low

Figure 4.2 Governance indicators (Azerbaijan).
Sources: Fund for Peace (2021), ND-GAIN (2020), UNDP (2020), World Bank (2021h, 2021i).

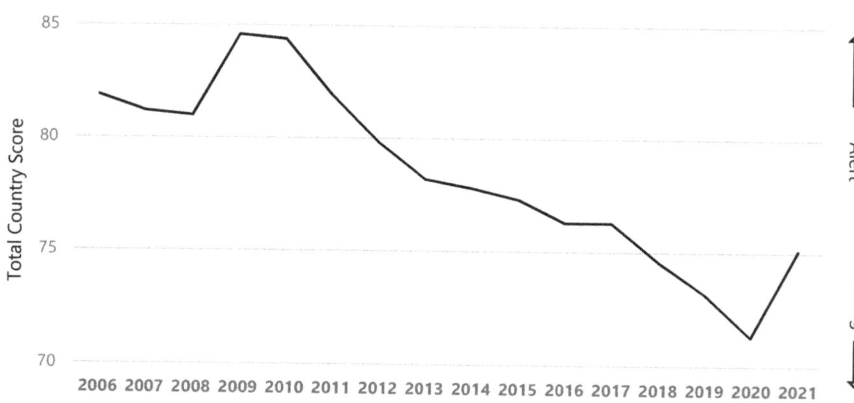

Figure 4.3 Azerbaijan's performance in the Fragile States Index 2006–2021.
Source: Fund for Peace (2021).

Note: Higher index numbers indicate a higher level of state fragility.

took place 'within a restrictive political environment and under laws that curtail fundamental rights and freedoms' and that 'observers reported widespread disregard for mandatory procedures, a lack of transparency and numerous serious irregularities' (European Union 2018).

The continuing, decades-old conflict with Armenia over the enclave/exclave of Nagorno-Karabakh constitutes a significant distinct security risk for Azerbaijan. While the exclave of Nakhchivan Autonomous Republic in Armenia belongs to Azerbaijan, the exclave of Nagorno-Karabakh lies wholly within Azerbaijan and possesses an Armenian ethnic majority. Nagorno-Karabakh is internationally recognised as part of Azerbaijan, but most of the region was actually governed by the Republic of Artsakh (formerly named Nagorno-Karabakh Republic) for several decades. Armenia and Azerbaijan had gone to war in 1988 over the Nagorno-Karabakh region, which resulted in a ceasefire in 1994 (CIA 2021). While this ceasefire remained precarious, it held – despite some intermittent clashes – until September 2020, when the conflict escalated into a full-fledged war again. Six weeks of fighting resulted in several thousand casualties (military and civilian) and eventually led to a new ceasefire mediated by Russia. In what from an Armenian view constituted a capitulation, Azerbaijan regained full control over large parts of Nagorno-Karabakh and several adjacent districts that had been controlled by Armenian forces. However, it had to accept the presence of Russian peacekeeping forces. Tensions in the region remain high and remain a risk for the development of the country (International Crisis Group 2020, 2021).

Azerbaijan faces relatively modest challenges with respect to climate change impacts, which are nonetheless not negligible. In particular, the agricultural sector of Azerbaijan is under threat because of increased evaporation, water shortages, and droughts due to rising temperatures (while the sea level of the Caspian Sea may in fact fall as a result of increased evaporation) (Azerbaijan 2015b: 68–76). Accordingly, Azerbaijan ranked 130 out of 180 countries on the Global Climate Risk Index for 2019 and 146 for the period 2000–2019. The Index assesses the extent to which countries have been subject to weather-related loss events (storms, floods, heat waves, etc.) (Germanwatch 2021). Considering eight life-supporting sectors (food, water, health, ecosystems, human habitat, and coastal, energy, and transportation infrastructure), the index of the Notre Dame Global Adaptation Initiative (ND-GAIN) for 2018 gives a medium score and rank to Azerbaijan for its overall vulnerability and readiness (ranked 73 out of 181 countries) (ND-GAIN 2020). Overall, climate change impacts are likely to be significant but less dramatic than in other countries.

Past and present efforts to decarbonise

Azerbaijan is a party to all major climate treaties. It ratified the UN Framework Convention on Climate Change (UNFCCC) in 1995, the Kyoto Protocol in 2000, and the Doha Amendment to the Kyoto Protocol in 2015. It also ratified the Paris Agreement in January 2017. It is a non-Annex I Party to the UNFCCC and considered a developing country for the purposes of the climate treaties.

Azerbaijan has submitted a relatively ambitious Nationally Determined Contribution (NDC) under the Paris Agreement. Accordingly, it aims at a GHG emission reduction of 35 percent by 2030 in comparison to 1990. This implies a reversal of the trend of rising emissions and a significant decrease from emission levels in the 2010s (see below and

Figure 4.4). The NDC lists general measures or areas of measures to achieve the set objectives for the energy sector, oil and gas extraction, residential and commercial sectors, the transport sector, the agricultural sector, the waste sector, and LULUCF. No further details (e.g., on timing or status of these measures) are provided (Azerbaijan 2015a). Future emissions and hence the ambitiousness of Azerbaijan's NDC depend heavily on economic development (which itself is highly dependent on oil and gas prices) and are consequently difficult to assess. According to one source, Azerbaijan's NDC does represent a certain level of ambition (Climatescope 2017; see also World Energy Council 2017). Nothing is known about Azerbaijan's intentions to strengthen its NDC.

According to the UNFCCC, Azerbaijan's GHG emissions excluding LULUCF (land use, land use change, and forestry) in 2013 (the latest available data) stood at 57,994.77 Gg CO_2 equivalent, a decline of about 20 percent compared to 1990 levels (73,385.78 Gt CO_2 equivalent) but an increase of about 43 percent since 2000. Azerbaijan's GHG emissions including LULUCF in 2013 stood at 51,796.66 Gg CO_2 equivalent compared to 69,695.78 Gg CO_2 equivalent in 1990 (minus 26 percent compared to 1990). The LULUCF sector and forestry in particular have thus consistently constituted a net carbon sink in Azerbaijan. As much as 85 percent of GHG emissions (excluding LULUCF) in 2013 were energy related, while agriculture accounted for about 11.5 percent (with industrial process figuring at 2.2 and waste at 1.3 percent). Energy industries, fugitive emissions (from oil and gas production), and transport accounted for the lion's share of energy-related emissions (UNFCCC 2013).

As Figure 4.4 indicates, GHG emissions data from other sources vary from those officially submitted to the UNFCCC and suggest that emissions have been on the rise since the end of the 1990s, potentially making the NDC even more ambitious. According to data of the World Resources Institute (WRI), emissions reached 80.26 Mt CO_2 equivalent in 2018, after a low of 56.94 Mt CO_2 equivalent in 1996, thereby almost attaining 1990 levels again. Azerbaijan's GHG emissions including LULUCF in 2018 stood at 78.35 Mt CO_2 equivalent and developed in a similar overall pattern (WRI 2021).

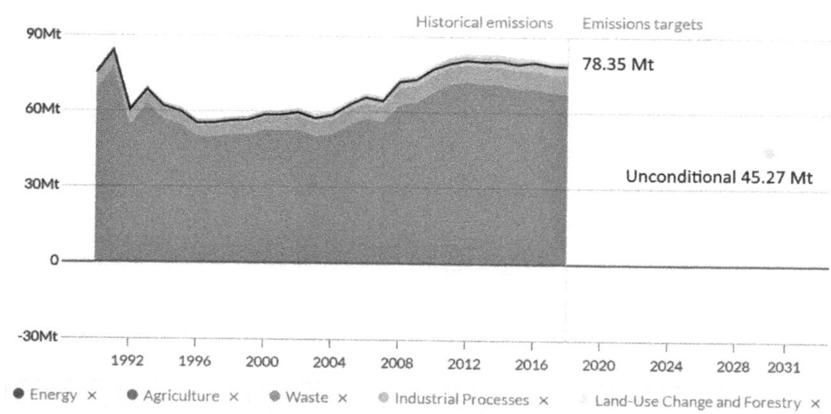

Figure 4.4 Azerbaijan's greenhouse gas emissions and emission targets (1990–2018).
Source: WRI (2021).

Contrasting the ambition of Azerbaijan's Paris NDC, its domestic climate policy framework is still emerging and in need of further development. The country has reported the adoption of a number of 'laws, state programs and regulatory acts concerning the Convention' and to address climate change in general, and certain institutional structures have been created (e.g., a ministry for the environment) (Azerbaijan 2015b: 10). However, these measures do not amount yet to a true climate policy framework that would need to address the country's main emission sectors, including oil and gas production, power production, transport, and agriculture. According to one assessment, Azerbaijan 'has no climate change policy, incentives or regulations' (Climatescope 2017).

Political framework conditions for the development and implementation of effective climate policy have been problematic. The oil and gas sector is closely intertwined with the government as the two core institutions in the oil and gas sector are state-run. First, the State Oil Company of the Azerbaijan Republic (SOCAR) engages in exploration, preparation, exploitation of oil and gas fields in Azerbaijan as well as transportation and processing and refining of oil and gas and related products. SOCAR accounts for about 20 percent of oil production in Azerbaijan, with the remaining 80 percent produced by international oil companies. SOCAR manages the country's two refineries, runs the country's pipeline system, and manages the country's oil and gas exports and imports. SOCAR is regulated directly by means of presidential regulations. It is overseen by the Energy Ministry that is also responsible for attracting foreign investment in the oil and gas sector as well as negotiations on pipelines and production-sharing agreements (EIA 2019; SOCAR 2021b).

Second, the State Oil Fund of the Republic of Azerbaijan (SOFAZ), created in 1999, accumulates savings from oil and gas revenues for the purpose of macroeconomic stabilisation of the country, to save resources for future generations, and to invest in national development projects, including through stimulating development of the non-oil and gas sector. SOFAZ remains much more closely related to and intertwined with the government than similar funds in other countries (such as the sovereign wealth fund in Norway). Hence, there are no limits on how much money can be transferred from the fund to the government so that it has in fact also served to balance budget deficits (see above). The president appoints the leadership of the fund directly, and SOFAZ' management only reports directly to the president (Osservatorio balcani 2012; Ibadoghlu 2019).

Trends and potential

Azerbaijan has longstanding intentions and efforts to diversify its economy beyond fossil fuels but has achieved only limited success so far. Education and training are crucial areas for developing its economy. Azerbaijan has also made significant investments to become a transport hub connecting Europe and Asia (including as part of the Chinese One Belt, One Road initiative). Renewable energy has considerable potential, but plans for its expansion have not come to fruition yet. While the response to the Covid-19 pandemic and resulting crisis has provided an opening for making progress on these fronts, with an emphasis on a green recovery, the available information does not indicate that Azerbaijan has so far made targeted efforts to exploit this opportunity (see https://platform2020redesign.org/countries/azerbaijan/, visited 13 July 2021).

Diversification of the economy

Longstanding efforts of Azerbaijan to diversify its economy, including through SOFAZ (see above), have only produced very limited results so far. Several programmes have been launched to support the diversification of the economy. They aim, among other things, to enhance the export potential of non-oil industry and more generally the development of new manufacturing, but also at developing the agricultural and service sectors of the economy. State investments have in particular served to support significant construction activity and the development of considerable transport infrastructure (including ports). Overall, the non-oil sector has seen significant growth in recent years, but the oil and gas sector remains crucial and its decline is in significant part a consequence of the falling oil price. Accordingly, international organisations such as the International Monetary Fund (IMF) and the World Bank have found that further efforts are needed (IMF 2016; Antidze 2018; see in general, Asgarova 2014; Guliyev 2015; Gurbanov et al. 2017; Lianlei and Baghirov 2016; UNECE 2021).

Transport-hub aspirations

One particular area of development is, based on the significant infrastructure investments of the past years, the role of Azerbaijan as a transport hub. Azerbaijan sees itself as a crucial connecting country between Europe and Asia. Beyond its position in the transport of oil and gas through pipelines, this concerns general air, train, and sea transport (see new Baku International Sea Trade Port and Baku-Tbilisi-Kars railway). Consequently, Azerbaijan has accepted China's offer to join its One Belt, One Road initiative that plays into Azerbaijan's aspiration to become a transport hub. Also beyond the One Belt, One Road initiative (and possibly driven by it), relations between Azerbaijan and China have grown closer over the past years, including trade, tourism, and even security and military cooperation. China has grown to become Azerbaijan's fourth biggest trading partner (Gotev 2017; Osmanli 2017; Azernews 2018a, 2018b, 2018c; Xinhuanet 2018; Jafarli 2020).

Renewable energy and clean energy

The significant potential for the expansion of renewable energy and clean energy investments in Azerbaijan still remains to be realised. Eight hydropower stations already supply about 5–6 percent of all electricity on average. However, solar and wind power are all but non-existent, supplying only a fraction of 1 percent of the total electricity consumption. As a result, renewable energy has a share of little more than 2 percent of total energy consumption (BP 2020; World Bank 2021a). Investments in renewable energy have been minimal, and the attractiveness of clean energy investments has remained low, despite considerable potential. Consequently, ambitious targets for the expansion of renewable energy by 2020 (20 percent of electricity consumption and 9.7 percent of overall energy consumption) have not been met (World Bank 2021a, 2021b; Climatescope 2017; UNECE 2017; EIA 2019; BP 2020).

Education and skills development

Azerbaijan's educational system is solid but has further potential for advancing excellence. It provides for a broad basic education and training of the population. Literacy is close to 100 percent and school life expectancy 14 years (CIA 2021). However, Azerbaijan ranked 68 out of the 78 countries participating in the Programme for International Student Assessment (PISA) in 2018 (ranked according to the average score in reading) (OECD 2019). The enrolment rate at universities (27.2 percent) has also remained low compared to other upper-middle-income countries (European Training Foundation 2020). Overall, Azerbaijan's workforce is considered medium-qualified, with 70 percent of the population holding at least an upper-secondary qualification and less than 10 percent described as low-skilled (graduating at most from compulsory education) (European Training Foundation 2016). Accordingly, Azerbaijan ranked 90 out of 146 countries in the 2019 Economic Complexity Index that measures 'the knowledge intensity of an economy' (Observatory of Economic Complexity 2021).

Azerbaijan has started several initiatives to improve its vocational education and training. In 2016, a Strategic Roadmap for vocational education and training was adopted, including several strategic goals such as the improvement of normative, legal, and economic conditions and the information base. For the years 2015–2020, a State Programme for the sector has been in place. Nevertheless, the supply of skilled graduates has not been sufficient to meet the demand (European Training Foundation 2016). In general, the improvement of conditions to facilitate the emergence of a knowledge-based economy has been one of the major focal points of Azerbaijan's efforts to diversify its economy and develop the non-oil sector (e.g., European Training Foundation 2016; Aliyev 2014: 13; European Training Foundation 2020; Suleymanov 2020).

Investments into education and training seem crucial for advancing the development and transition of the country and the diversification of its economic base. In 2020, 35.5 percent of the workforce was still employed in the agricultural sector (a slight decline from 38.2 percent in 2010), nearly 15 percent in industry, and nearly 50 percent in the service sector (Statista 2021). Especially the relatively large share of employment in the agricultural sector, which provides for more than a third of employment but only contributes 7 percent of GDP, constitutes a challenge. Against the backdrop of a growing population of young people entering the labour market, new employment opportunities will need to be created in the private sector (World Bank 2015; European Training Foundation 2016, 2020). Whereas Azerbaijan ranked 58 out of 141 countries on the World Economic Forum's Global Competitiveness Index in 2019, the Forum has identified education and training among the most important areas with potential and need for significant improvement (World Economic Forum 2019).

EU-Azerbaijan cooperation

EU-Azerbaijan cooperation can build on a firm and broad institutional framework. The EU is Azerbaijan's most important trading partner, accounting for nearly half of the latter's exports. Energy has been key to EU-Azerbaijan relations, with a strong focus on oil and gas. Renewable energy has received less attention, and existing cooperation on education provides a basis for further development.

Institutional framework: Partnership and Cooperation Agreement, European Neighbourhood Policy, and Eastern Partnership

Cooperation between the EU and Azerbaijan can rely on a firm institutional basis. First of all, a Partnership and Cooperation Agreement has been in force since 1999 and addresses political dialogue, trade, investment, economic matters, legislation, and culture. In 2017, negotiations on a new comprehensive agreement that is to replace the Partnership and Cooperation Agreement of 1999 were launched. The new agreement should 'offer a renewed basis for political dialogue and mutually beneficial cooperation between the EU and Azerbaijan' (EEAS 2017a). In this context, it is noteworthy that a survey conducted in 2017 found that pro-EU feelings have risen in Azerbaijan, with 47 percent of the respondents having a positive image of the EU and 68 percent assessing Azerbaijan's relations with the EU as good (EEAS 2017b).

Azerbaijan is also part of the European Neighbourhood Policy (ENP, since 2004) and the Eastern Partnership (since its inception in 2009). These frameworks provide a basis for cooperation on democracy, the rule of law, prosperity, and social cohesion, including cooperation in the domain of environmental policy. In this context, Azerbaijan has received support from and has had access to a number of financial support mechanisms, including the European Neighbourhood Instrument (with an allocation of up to €169 million for 2014–2020). The Neighbourhood Instrument's support has focused on regional and rural development, justice sector reform, education and skills development, and general capacity-building and institution-building (EEAS 2018). The EU Clima East and EU4Climate projects have assisted Azerbaijan in the implementation of the Paris Agreement, including the preparation and development of its (Intended) Nationally Determined Contribution and the formulation of a national climate change strategy (as seen above, to limited effect so far) (EU4Climate 2021).

Trade

The EU trade relationship with Azerbaijan has evolved significantly in the 2000s. The Partnership and Cooperation Agreement did not include tariff preferences but abolished a number of trade quotas between the EU and Azerbaijan and aimed at Azerbaijan adapting to several European standards. Until 2014, Azerbaijan benefitted from the EU Generalised Scheme of Preferences that reduces EU import duties by about 66 percent. Azerbaijan lost this benefit in 2014 when it was classified as an upper-middle-income country by the World Bank for three years in a row. Azerbaijan applied to become a member of the World Trade Organization (WTO) in 1997, with negotiations still ongoing. The EU supports Azerbaijan in its ambition to join the WTO.

The EU is Azerbaijan's most important trading partner, accounting for more than a third of Azerbaijan's total trade in 2020: about half of exports went to the EU and more than 20 percent of Azerbaijan's imports came from the EU in 2020 (a particular year due to the Covid-19 pandemic since trade volumes contracted by more than a third from 2019 to 2020). EU imports from Azerbaijan (overwhelmingly oil and gas – see below) were worth €10.6 billion in 2019 and €6.4 billion in 2020, and total trade amounted to €12.4 billion in 2019 and €7.9 billion in 2020 (European Commission 2021; see also Chatham House 2021). EU exports to Azerbaijan are mainly machinery, transport equipment, manufactured goods, and chemicals (European Commission 2021).

Energy: focus on fossil fuels

The EU is the top destination of Azeri fossil fuel exports, accounting for about half in 2019 (European Commission 2021; Chatham House 2021). EU imports from Azerbaijan currently mainly consist of oil, with increased gas exports planned for the future (from 2020). Oil and gas account for about 98 percent of EU imports from Azerbaijan. The main destinations for Azeri gas exports are Turkey (by far the biggest client), Georgia, and Greece (Razayeva 2015; BP 2020) – with the EU set to increase its share in the wake of the Southern Gas Corridor becoming operational in late 2020 (see also Figure 4.1).

As obvious from the trade figures, energy has been, and has emerged as, a major area of the relationship between Azerbaijan and the EU. In 2006, the Memorandum of Understanding on a Strategic Partnership between the Republic of Azerbaijan and the European Union in the Field of Energy was signed (covering harmonisation of legislation, enhancing security of supply and transit systems, development of renewable energy systems, and technical cooperation). In 2016, furthermore, the *EU4Energy* initiative was launched that supports Eastern Partnership countries in achieving their energy policy objectives. In 2019, Azerbaijan also joined the Eastern Europe Energy Efficiency and Environment Partnership (EEAS 2017c; European Council 2020; EU4Climate 2021). Within this broader framework, a strategic emphasis has been put on developing the Southern Gas Corridor in order to bring Azeri gas to Europe. In 2011, then Commission President Barroso and Azerbaijan's president Aliyev signed a Joint Declaration on the Southern Gas Corridor. The Gas Corridor was defined as an EU Project of Common Interest. On this basis, it has received active support from the European Commission and, in February 2018, a loan of €1.5 billion from the European Investment Bank (EIB). The apparent focus of the relationship on developing fossil fuel trade, and especially gas deliveries from Azerbaijan to Europe (now via the Southern Gas Corridor), has drawn significant criticism because of its potential conflict with medium- and long-term climate objectives (Teffer 2018).

Education: room for further development

In addition to trade and energy, EU-Azerbaijan relations also cover education and training, and mobility. Azerbaijan is part of the Erasmus+ programme that supports higher education institutions, staff, and students. From 2015 to 2019, the EU funded nearly 1,600 students and staff member mobility from Azerbaijan to Europe and 900 European students and staff mobility to Azerbaijan. About 3,300 young people and youth workers from Azerbaijan participated in different types of short-term stays in Europe. The EU, several of its member states, and Azerbaijan also operate under a Mobility Partnership that was signed in 2013. Perhaps most notable, a Visa Facilitation Agreement that entered into force in 2014 makes it easier for Azerbaijani citizens to acquire travel visas for EU member states (European Council 2020).

Security and geopolitics

Finally, EU-Azerbaijan relations have a significant security and geopolitical component. Geopolitically, the ambiguous position of Azerbaijan vis-à-vis Russia is significant. Russia has acted as an important protector of Armenia, while at the same time fostering close relations to Azerbaijan (as a member of the Commonwealth of Independent States). This has not least enabled Russia to effectively mediate between Armenia and Azerbaijan in the war over Nagorno-Karabakh in 2020. At the same time, Azerbaijan remains interested in developing its relations beyond Russia. As a result, Azerbaijan's relations with China have grown over the past years, including with respect to the economy and security (Azernews 2018a, 2018b; Shabhazov 2018; Xinhuanet 2018; Jafarli 2020).

The EU has an opportunity to redefine and fill its place in the regional security architecture. The EU has for a long time supported efforts by the Co-Chairs of the OSCE Minsk group to resolve the conflict and has generally promoted relevant confidence- and peace-building activities, including through the *European Partnership for the peaceful settlement of the conflict over Nagorno-Karabakh* (EEAS 2017a; The European Partnership for the Peaceful Settlement of the Conflict over Nagorno-Karabakh 2018). Also, Azerbaijan sought closer cooperation in the field of security policy, including a more proactive role of the EU in the Nagorno-Karabakh conflict (Merabishvili 2015). The EU can build on these previous efforts and adapt them to the changed context after the 2020 war, which has not reduced the need for a lasting and stable settlement of the conflict.

Conclusions

Azerbaijan is highly dependent on oil and gas exports for its economic welfare and political stability. It is located in the Caucasian region that is characterised by a number of conflicts and a fragile, and evolving, geopolitical balance, involving Russia, China, the United States, and the EU. The EU is one of the main export markets for oil and gas from Azerbaijan (while Azerbaijan's relations with China are becoming more important), and Azerbaijan has particular relevance for the EU as a neighbourhood country.

Without proper accompanying political efforts, decarbonisation of Europe and the world has the potential to undermine political and economic stability of Azerbaijan. Phasing out fossil fuel production and exports will mean replacing Azerbaijan's main engine of economic development and source of government income, which could reinforce existing risks to political stability, including limited political freedom and military conflicts that remain virulent and await a lasting settlement, in particular with respect to Nagorno-Karabakh. Decarbonisation hence poses the challenge to find alternatives to oil and gas as a basis for Azerbaijan's economic well-being and political settlement in a delicate and challenging broader political context.

There are a number of areas of mutual interest between Azerbaijan and the EU beyond oil and gas that can provide a basis for fruitfully developing the relationship under decarbonisation. At a geopolitical level, the EU has an interest in developing this relationship because Azerbaijan belongs to its neighbourhood, while Azerbaijan has an interest in diversifying its external relations and opportunities. While Russia's role in the region remains prominent, it is also precarious in its relations within the region given specific own interests. More specifically with respect to the decarbonisation challenge, Azerbaijan has an interest in developing its economy beyond oil and gas, which the EU can assist in developing in important ways. This will, however, require a reorientation away from the established focus on the fossil fuel relationship (including the Southern Gas Corridor).

Initiating an evolution of the EU-Azerbaijan relationship under decarbonisation will likely require a step-by-step approach, carefully considering and balancing economic and broader geopolitical aspects under the climate imperative.

On the basis of the preceding analysis, several prominent and promising focal areas emerge for developing the relationship between the EU and Azerbaijan beyond oil and gas:

- **Education and training** are a central part of the effort at economic diversification and developing a knowledge-based economy. Building on existing cooperation (e.g., in the context of the EU Erasmus+ programme, etc.), bilateral cooperation can be intensified so as to bring the significant resources of the EU and its member states to bear to assist in enhancing and shaping education and training in Azerbaijan. One significant focus in this respect could be cooperation supporting the expansion of renewable energy (e.g., a degree in renewable energy in engineering).
- **The further development of the energy system, especially the expansion of renewable energy** promises obvious advantages by making the economy less dependent on oil and gas and developing energy infrastructure. While it is urgently required for moving toward decarbonisation, it also makes sense with continued exploitation of domestic oil and gas resources in the medium term as it will increase the share of (declining) resources available for export. Concrete first steps may include studying the risk of stranded assets with respect to relevant fossil fuel investments and the creation of a policy dialogue on the potential of clean/renewable energy.
- **Strengthening the rule of law and advancing the fight against corruption** is one of the crucial bases of attracting foreign investors at a large scale, including SMEs. The EU and its member states have significant experience and expertise to offer to this end. Part of such a focus could also be a dialogue on strengthening financial governance, including SOFAZ and the changing landscape of climate-proofing investments.
- Multiple other areas can be further developed over time, including water management and advancing mutual market access and investments. Also, an enhanced role of the EU in managing the conflict between Armenia and Azerbaijan concerning Nagorno-Karabakh deserves particular attention.

Overall, there is hence a range of options for developing the EU-Azerbaijan relationship beyond oil and gas. Such a renewed direction for the bilateral relationship would promise significant returns for Azerbaijan as it would help address stability risks arising from the dependence of the country on oil and gas and put Azerbaijan's economic development on a broader basis, preparing for a medium-term decline of returns from decreasing oil and gas reserves. It would at the same time help put EU-Azerbaijan relations on a broader basis making them fit for a decarbonising world and strengthening them in a precarious geopolitical constellation.

Note

1 Production, consumption, and export data vary between different sources (including IEA, CIA, BP Statistical Review) but all give roughly the same overall picture.

References

Aliyev, I. (2014) 'State Program on the Development of Industry in the Republic of Azerbaijan in 2015-2020'. Available at: http://extwprlegs1.fao.org/docs/pdf/aze175121.pdf (accessed 20 November 2019).

Antidze, M. (2018) 'Development banks tell Azerbaijan to diversify economy more', *Reuters*, 21 February 2018. Retrieved 20 December 2018 from https://www.reuters.com/article/azerbaijan-economy/development-banks-tell-azerbaijan-to-diversify-economy-more-idUKL8N1QB3KX?edition-redirect=uk.

Asgarova, N. (2014) 'More Diversified and Sustainable Economic Growth in the Next Few Decades'. Retrieved 20 December 2018 from https://www.ebrd.com/documents/comms-and-bis/more-diversified-and-sustainable-economic-growth-in-the-next-few-decades.pdf.

Azerbaijan (2015a) 'Intended Nationally Determined Contribution'. Available at: https://www4.unfccc.int/sites/ndcstaging/PublishedDocuments/Azerbaijan%20First/INDC%20Azerbaijan.pdf (accessed 20 November 2019).

Azerbaijan (2015b) 'Third National Communication to the United Nations Framework Convention on Climate Change Republic of Azerbaijan'. Available at: https://unfccc.int/resource/docs/natc/azenc3.pdf (accessed 20 December 2018).

Azernews (2018a) 'Azerbaijan Becomes Significant Transport Hub for Europe'. Retrieved 20 December 2018 from https://www.azernews.az/business/128782.html.

Azernews (2018b) 'Azerbaijan Opens Trade Representative Office in China'. Retrieved 20 December 2018 from https://www.azernews.az/business/126688.html.

Azernews (2018c) 'Azerbaijan, China Eye Further Military Cooperation'. Retrieved 20 December 2018 from https://www.azernews.az/nation/131274.html.

Bertelsmann Stiftung (2020) *Bertelsmann Transformation Index. Azerbaijan Country Report 2020*. Available at: https://www.bti-project.org/content/en/downloads/reports/country_report_2020_AZE.pdf (accessed 16 June 2021).

Bertelsmann Stiftung (2021) *Sustainable Development Report 2021. The Decade of Action for the Sustainable Development Goals*. Available at: https://s3.amazonaws.com/sustainabledevelopment.report/2021/2021-sustainable-development-report.pdf (accessed 16 June 2021).

BP (2020) *BP Statistical Review of World Energy 2020*. Available at: https://www.bp.com/content/dam/bp/business-sites/en/global/corporate/pdfs/energy-economics/statistical-review/bp-stats-review-2020-full-report.pdf (accessed 16 June 2021).

Breban, D. and Mukhtarov, E. (2017) *The 2018 State Budget Project of Azerbaijan: From Consolidation to High Spending?* Available at: http://cesd.az/new/wp-content/uploads/2017/11/CESD-Research-2018-Budget-Projections.pdf (accessed 20 December 2018).

Chatham House (2021) 'Resource Trade. Earth'. Retrieved 16 June 2021 from https://resourcetrade.earth/data?year=2017&exporter=31&importer=97&category=4&units=value.

CIA (2021) 'The World Factbook. Azerbaijan'. Retrieved 16 June 2021 from https://www.cia.gov/the-world-factbook/countries/azerbaijan/.

Climatescope (2017) 'Climate Scope 2017. Which Emerging Market Is Most Attractive for Clean Energy Investment?' Retrieved 20 December 2018 from http://2017.global-climatescope.org/en/.

EEAS – European Union External Action Service (2017a) *EU-Azerbaijan Relations. Factsheet*. Available at: https://eeas.europa.eu/sites/default/files/eap_factsheet_azerbaijan_en.pdf (accessed 20 December 2018).

EEAS (2017b) '2017 Public Opinion Poll about the European Union in Azerbaijan'. Retrieved 20 December 2018 from https://eeas.europa.eu/delegations/azerbaijan_cs/33458/2017%20Public%20Opinion%20Poll%20about%20the%20European%20Union%20in%20Azerbaijan.

EEAS (2017c) 'European Union, Energy Community and the Ministry of Energy of Georgia Mark the Opening of EU4ENERGY Regional Office in Tbilisi to Support Energy Reforms in Georgia, Armenia and Azerbaijan'. Retrieved 20 November 2019 from https://eeas.europa.eu/delegations/georgia_kk/22479/European%20Union, %20Energy%20Community%20and%20the%20Ministry%20of%20Energy%20of%20Georgia%20mark%20the%20opening%20of%20EU4ENERGY%20regional%20office%20in%20Tbilisi%20to%20support%20energy%20reforms%20in%20Georgia, %20Armenia%20and%20Azerbaijan.

EEAS (2018) 'Programming of the European Neighbourhood Instrument (ENI) – 2014-2020. Single Support Framework for EU Support to Azerbaijan (2014-2017)'. Available at: http://eeas.europa.eu/archives/docs/enp/pdf/financing-the-enp/azerbaijan_2014_2017_programming_document_en.pdf (accessed 20 November 2019).

EIA – US Energy Information Administration (2019) 'Country Analysis Executive Summary: Azerbaijan' and 'Background Reference: Azerbaijan'. Retrieved 13 July 2021 from https://www.eia.gov/international/analysis/country/AZE.

EIA – US Energy Information Administration (2021) 'Proved Reserves of Natural Gas'. Retrieved 16 June 2021 from https://www.eia.gov/international/data/world.

EU4Climate (2021) 'Azerbaijan'. Retrieved 13 July 2021 from https://eu4climate.eu/azerbaijan/.

European Commission (2021) 'European Union, Trade in Goods with Azerbaijan'. Retrieved 13 July 2021 from https://webgate.ec.europa.eu/isdb_results/factsheets/country/details_azerbaijan_en.pdf.

European Council (2020) 'EU Relations with Azerbaijan'. Retrieved 17 June 2021 from http://www.consilium.europa.eu/en/policies/eastern-partnership/azerbaijan/.

European Parliament DG for External Policies Policy Department (2017) *EU Relations with Armenia and Azerbaijan*. Available at: http://www.europarl.europa.eu/RegData/etudes/IDAN/2017/603846/EXPO_IDA(2017)603846_EN.pdf (accessed 20 December 2018).

European Training Foundation (2016) *Azerbaijan. Country Strategy Paper 2017-20*. Available at: https://www.etf.europa.eu/sites/default/files/document/Country%20Strategy%20Paper%202017-20%20Actions%202020_Azerbaijan.pdf (accessed 20 November 2019).

European Training Foundation (2020) *Azerbaijan. Education, Training and Employment Developments 2020*. Available at: https://www.etf.europa.eu/sites/default/files/document/Country%20Fiche%202020%20Azerbaijan%20-%20Education_%20Training%20and%20Employment%20Developments.pdf (accessed 20 July 2021).

European Union (2018) 'Statement by the Spokesperson on the Presidential Elections in the Republic of Azerbaijan'. Retrieved 20 December 2018 from https://eeas.europa.eu/delegations/azerbaijan/42873/statement-spokesperson-presidential-elections-republic-azerbaijan_en.

Freedom House (2021) 'Azerbaijan Profile'. Retrieved 17 June 2021 from https://freedomhouse.org/country/azerbaijan/freedom-world/2021.

Fund for Peace (2021) 'Fragile States Index 2021'. Retrieved 26 July 2021 from https://fragilestatesindex.org/.

Germanwatch (2021) 'Global Climate Risk Index 2021'. Retrieved 17 June 2021 from https://germanwatch.org/en/19777.

Gotev, G. (2017) 'Azerbaijan moves forward with "5 star" free trade hub', *Euractiv*, 08 April 2017. Retrieved 20 December 2018 from https://www.euractiv.com/section/transport/news/azerbaijan-moves-forward-with-5-star-free-trade-hub/.

Guliyev, F. (2015) '"After Us, the Deluge": Oil Windfalls, State Elites and the Elusive Quest for Economic Diversification in Azerbaijan. Caucasus Analytical Digest, Nr. 69'. Available at SSRN: https://ssrn.com/abstract=2559800.

Gurbanov, S., Nugent, J.B., and Mikayilov, J. (2017) 'Management of oil revenues: Has that of Azerbaijan been prudent?', *Economies*, 5(2), pp. 19. https://doi.org/10.3390/economies5020019.

Ibadoghlu, G. (2019) 'Retrospective Assessment of the State Oil Fund Performance'. Available at SSRN: https://ssrn.com/abstract=3489640 or http://dx.doi.org/10.2139/ssrn.3489640 (accessed 20 July 2021).

Indexmundi (2020) 'Population Below Poverty Line'. Retrieved 16 June 2021 from https://www.indexmundi.com/g/g.aspx?c=aj&v=69.

IEA – International Energy Agency (2020) 'Energy Subsidies'. Retrieved 2 July 2021 from https://www.iea.org/weo/energysubsidies/.

IMF – International Monetary Fund (2016) 'Azerbaijan's Opportunity to Reboot, Diversify Economy'. Retrieved 20 December 2018 from https://www.imf.org/en/News/Articles/2016/09/15/NA091516-Azerbaijan-Opportunity-to-Reboot-Diversify-Economy.

International Crisis Group (2020) *Getting from Ceasefire to Peace in Nagorno-Karabakh*, 10 November 2020. Available at: https://reliefweb.int/sites/reliefweb.int/files/resources/Getting%20from%20Ceasefire%20to%20Peace%20in%20Nagorno-Karabakh%20_%20Crisis%20Group.pdf (accessed 13 July 2021).

International Crisis Group (2021) 'The Nagorno-Karabakh Conflict: A Visual Explainer'. Retrieved 13 July 2021 from https://www.crisisgroup.org/content/nagorno-karabakh-conflict-visual-explainer.

Jafarli, S. (2020) 'Azeri-China Relations'. Available at: https://bakuresearchinstitute.org/en/azerbaijan-china-relations/ (accessed 21 July 2021).

Lianlei, B. and Baghirov, O. (2016) 'Economic Diversification in Azerbaijan'. China Institute for International Studies, 13 September 2016. Retrieved 20 December 2018 from http://www.ciis.org.cn/english/2016-09/13/content_9030607.htm.

Merabishvili, G. (2015) 'The EU and Azerbaijan: Game on for a more normative policy?' CEPS Policy Brief, March 2015. Retrieved 24 October 2021 from https://web.archive.org/web/20180427014516id_/http://aei.pitt.edu/63166/1/PB329_EU_Policy_towards_Azerbaijan_G_Merabishvili.pdf.

ND-GAIN – Notre Dame Global Adaptation Initiative (2020) 'ND-GAIN Country Index'. South Bend, IN: University of Notre Dame. Retrieved 05 July 2021 from https://gain.nd.edu/our-work/country-index/.

Observatory of Economic Complexity (2021) 'Azerbaijan'. Retrieved 13 July 2021 from https://oec.world/en/profile/country/aze.

OECD – Organisation for Economic Co-operation and Development (2019) *PISA 2018 Results. Combined Executive Summaries Volume I, II & III*. Available at: https://www.oecd.org/pisa/Combined_Executive_Summaries_PISA_2018.pdf (accessed 14 June 2021).

Osmanli, C. (2017) 'Azerbaijan: A new energy and transport hub between Europe and Asia', *The Parliament Magazine*, 24 November 2017. Retrieved 20 December 2018 from https://www.theparliamentmagazine.eu/blog/azerbaijan-new-energy-and-transport-hub-between-europe-and-asia.

Osservatorio balcani e caucaso transeuropa (2012) 'Azerbaijan's Oil Dependence'. Retrieved 20 December 2018 from https://www.balcanicaucaso.org/eng/Areas/Azerbaijan/Azerbaijan-s-oil-dependence-123328.

Razayeva, G. (2015) *The Outlook for Azerbaijani Gas Supplies to Europe: Challenges and Perspectives*. The Oxford Institute for Energy Studies OIES Paper NG 97. Available at: https://www.oxfordenergy.org/wpcms/wp-content/uploads/2015/06/NG-97.pdf (accessed 20 December 2018).

Santander (2018) 'Azerbaijan: Foreign Investment'. Retrieved 20 December 2018 from https://en.portal.santandertrade.com/establish-overseas/azerbaijan/investing?&actualiser_id_banque=oui&id_banque=0&memoriser_choix=memoriser.

Shepard, W. (2016) 'What Azerbaijan plans to do when the oil runs out', *Forbes*, 3 December 2016. Retrieved 20 December 2018 from https://www.forbes.com/sites/wadeshepard/2016/12/03/what-azerbaijan-plans-to-do-when-the-oil-runs-out/#706ed4063780.

SOCAR (2021a) 'Economics and Statistics. Gas Production (Annual)'. Retrieved 13 July 2021 from http://www.socar.az/socar/en/economics-and-statistics/economics-and-statistics/gas-production.

SOCAR (2021b) 'About SOCAR'. Retrieved 13 July 2020 from https://socar.de/en/socar-2/2214-2/.

Statista (2021) 'Azerbaijan: Distribution of Employment by Economic Sector from 2010 to 2020'. Retrieved 13 July 2021 from https://www.statista.com/statistics/457598/employment-by-economic-sector-in-azerbaijan/.

Suleymanov, T. (2020) 'Transformation of higher education in Azerbaijan: Reforms, policies and current trends', *The Journal of Economic Sciences: Theory and Practice*, 77(2), pp. 40–60.

Shabhazov, F. (2018) 'China to Europe by Azerbaijan's Trans-Caspian Gateway', *The Diplomat*, 16 February 2018. Retrieved 20 December 2018 from https://thediplomat.com/2018/02/china-to-europe-by-way-of-azerbaijans-trans-caspian-gateway/.

Teffer, P. (2018) 'EU bank courts controversy with €1.5bn gas pipe investment', *EUObserver*, 09 February 2018. Retrieved 20 December 2018 from https://euobserver.com/energy/140926.

The European Partnership for the Peaceful Settlement of the Conflict over Nagorno-Karabakh (2018) 'Home'. Retrieved 20 December 2018 from http://www.epnk.org/.

Trading Economics (2021a) 'Azerbaijan Unemployment Rate'. Retrieved 16 June 2021 from https://tradingeconomics.com/azerbaijan/unemployment-rate.

Trading Economics (2021b) 'Azerbaijan Government Debt to GDP'. Retrieved 16 June 2021 from https://tradingeconomics.com/azerbaijan/government-debt-to-gdp.

Transparency International (2021) 'Corruption Perceptions Index. Azerbaijan'. Retrieved 16 June 2021 from https://www.transparency.org/en/cpi/2020/index/nzl.

UNDP – United Nations Development Programme (2020) 'Human Development Report'. Available at: http://hdr.undp.org/en/countries (accessed 26 July 2021).

UNECE – United Nations Economic Commission for Europe (2017) *Renewable Energy Status Report*. Available at: https://www.unece.org/fileadmin/DAM/energy/se/pp/renew/Renewable_energy_report_2017_web.pdf (accessed 20 December 2018).

UNECE (2021) 'Sub-regional Innovation Policy Outlook 2020: Eastern Europe and the South Caucasus'. Available at: https://unece.org/sites/default/files/2021-06/UNECE_Sub-regional_IPO_2020_Publication.pdf.

UNFCCC – United Nations Framework Convention on Climate Change (2013) 'Greenhouse Gas Inventory Data – Detailed Data by Party: Azerbaijan'. Retrieved 17 June 2021 from http://di.unfccc.int/detailed_data_by_party.

World Bank (2015) 'The Jobs Challenge in the South Caucasus – Azerbaijan'. Retrieved 09 July 2021 from https://www.worldbank.org/en/news/feature/2015/01/06/jobs-challenge-in-the-south-caucasus-azerbaijan.

World Bank (2021a) 'Renewable Energy Consumption (% of Final Energy Consumption)'. Retrieved 16 June 2021 from https://data.worldbank.org/indicator/EG.FEC.RNEW.ZS?locations=AZ.

World Bank (2021b) 'Renewable Electricity Output (% of Total Electricity Output)'. Retrieved 17 June 2021 from https://data.worldbank.org/indicator/EG.ELC.RNEW.ZS?locations=AZ&view=chart.

World Bank (2021c) 'GDP (Current US$)'. Retrieved 16 June 2021 from https://data.worldbank.org/indicator/NY.GDP.MKTP.CD?locations=AZ.

World Bank (2021d) 'GDP Per Capita (Current US$)'. Retrieved 16 June 2021 from https://data.worldbank.org/indicator/ny.gdp.pcap.cd?locations=AZ.

World Bank (2021e) 'GDP (Constant 2010 US$)'. Retrieved 16 June 2021 from https://data.worldbank.org/indicator/NY.GDP.MKTP.KD?locations=AZ.

World Bank (2021f) 'GDP Per Capita (Constant 2010 US$)'. Retrieved 16 June 2021 from https://data.worldbank.org/indicator/NY.GDP.PCAP.KD?locations=AZ.

World Bank (2021g) 'Fuel Exports (% of Merchandise Exports)'. Retrieved 15 July 2021 from https://data.worldbank.org/indicator/TX.VAL.FUEL.ZS.UN?locations=AZ.

World Bank (2021h) 'Worldwide Governance Indicators 2019'. Retrieved 26 July 2021 from http://info.worldbank.org/governance/wgi/#reports.

World Bank (2021i) *Regulatory Indicators for Sustainable Energy (RISE)*. Retrieved 26 July 2021 from https://rise.esmap.org.

World Economic Forum (2018) *The Inclusive Development Index 2018. Summary and Data Highlights.* Available at: http://www3.weforum.org/docs/WEF_Forum_IncGrwth_2018.pdf (accessed 20 December 2018).

World Economic Forum (2019) 'The Global Competitiveness Report 2019'. Cologny/Geneva: World Economic Forum. Available at http://www3.weforum.org/docs/WEF_TheGlobalCompetitivenessReport2019.pdf.

World Energy Council (2017) 'Energy Trilemma Index'. Retrieved 28 February 2018 from https://trilemma.worldenergy.org/.

WRI – World Resources Institute (2021) 'Climate Watch Azerbaijan: Greenhouse Gas Emissions and Emissions Targets'. Retrieved 17 June 2021 from https://www.climatewatchdata.org/countries/AZE.

Xinhuanet (2018) 'Interview: Azerbaijan-China Ties Continue to Strengthen, Says Azerbaijani MP'. Retrieved 20 December 2018 from http://www.xinhuanet.com/english/2018-04/05/c_137090333.htm.

5 Colombia

The double challenge of internal pacification and decarbonisation

Introduction

The Republic of Colombia (henceforth Colombia) is a country with varied geography and climate, ranging from the cooler climate of the central Andean highlands to a tropical climate along the coast, in the Llanos, its vast eastern plains, and in the Amazon rainforest to the south. Most people therefore live in the north and west of the country where most of the country's agricultural activities and natural resources are found (CIA 2021). Colombia is largely urbanised with around 82 percent of the population living in cities (CIA 2021) and Bogotá, the world's fourth highest capital city, has a population of 7.6 million. With 51.3 million inhabitants, Colombia has the fourth largest population in Latin America (World Population Review 2021; see also DANE 2018; CIA 2021).

Colombia is at an important turning point in its political, economic, and social development. In late 2016, the former coalition government led by President Juan Manuel Santos completed six years of negotiations to sign a peace agreement (Government of Colombia 2015) with the country's largest leftist armed militant group, the Revolutionary Armed Forces of Colombia (known by its Spanish acronym, FARC). This ended a long and complex conflict that had spanned over 50 years, leaving over 220,000 people dead, 25,000 disappeared, and 5.7 million displaced (Felter and Renwick 2017). The success of the peace agreement rests on continued commitment from both sides and major investment and reform in rural areas by the Colombian government. As well as opening up new opportunities for development, the peace process created new dynamics and challenges for the government, headed since August 2018 by President Iván Duque Márquez, known as Duque, who has been unsupportive of the peace process. Reinforced by the Covid-19 pandemic, the peace agreement has been increasingly fragile and has been found to be at the verge of collapse (Fund for Peace 2020; Reith 2021).

In the decade up to 2015, the Colombian economy grew strongly by almost 5 percent per year. High commodity prices largely protected it from the impacts of the 2008 financial crisis, and it has proved more resilient to falls in the price of crude oil between 2014 and 2016 than other oil-producing countries (OECD 2017). The proportion of people living with less than US$1.90 per day fell from 21.1 percent in 2001 to 4.9 percent in 2019 (World Bank 2021a). According to national poverty lines, however, about 36 percent of the population was living in poverty in 2019 (World Bank 2021b). In May 2018, Colombia became a member of the Organisation for Economic Co-operation and Development

DOI: 10.4324/9781003183037-5

(OECD), after major reforms to bring its labour policies, justice system, corporate governance of state-owned enterprises, and trade rules among others into line with OECD standards (OECD 2018a). However, Colombia's unemployment rates are still higher than the OECD average (OECD 2018b), and inequality remains high due to the large disparities in income between rural and urban areas. The Covid-19 pandemic has aggravated the remaining challenges, although prospects for a strong recovery exist (OECD 2021).

Colombia has many natural resources, including petroleum, coal, and natural gas, as well as precious metals and minerals, such as gold, copper, iron ore, and nickel (CIA 2021). Although the economy is well diversified, its extractive sectors remain important. The country also has rich, fertile land, and the agricultural sector could grow rapidly with an effective implementation of the peace process (Oxford Business Group 2019). This could diversify exports away from fossil fuels. However, the sector is highly vulnerable to climate change impacts and may also increase emissions by driving deforestation. Colombia is among the world's mega-diverse nations. Although only covering 1 percent of the Earth's surface, it is home to 10 percent of all known species and has forests covering 55 percent of its land area (OECD 2014; UNEP 2016).

In parallel with the other case studies, this chapter will explore the challenges and opportunities Colombia and Colombia-EU relations are facing under decarbonisation in five steps. The next section first substantiates the dependence of Colombia on the production and export of coal and oil and hence the country's vulnerability to price fluctuations and the phase-out of fossil fuels. It also analyses other fragility and security risks, most importantly the decades-long internal conflicts with the FARC (and other militant groups). Subsequently, the status of Colombia's climate policy is assessed, which is comparatively advanced. This is followed by the review of relevant trends and potentials, especially regarding efforts to diversify the economy in different directions that have already borne some fruit. The one but last section reviews the framework and basis of EU-Colombia relations beyond fossil fuels. The final section concludes and identifies focal points for EU foreign policy towards Colombia under decarbonisation.

Exposure and risks

Exposure to global decarbonisation trends

Colombia is a significant producer of crude oil and coal. A large proportion of production is destined for export, generating revenues that are important to the economy. With oil reserves diminishing, the future of the Colombian oil sector is uncertain. Oil rents have also been vulnerable to international price fluctuations, although the Colombian economy fared better than those of other oil producers during the 2014–2016 oil price slump. Decarbonisation is contributing to shifts in Colombia's coal export markets, leading it to become increasingly reliant on just a few markets.

Although not one of the world's leading energy producers, Colombia is a significant producer of oil, coal, and gas in South America. It is the third largest oil producer in the region, and between 2009 and 2019 overall production rose 32 percent to around 886,000 barrels per day (BP 2020). However, production decreased 12 percent from 2015 to 2018 (BP 2020), in part due to falling oil sector investment, as well as frequent outages due to attacks by a Marxist insurgent group, the National Liberation Army (ELN) (Acosta and Cobb 2017). Colombia is now by far the largest coal producer in Latin America and the eighth largest in the world, with 1.4 percent of total world production in 2019. That year Colombian coal production stood at 56.6 million tonnes of oil equivalent, a 13.4 percent increase on production levels in 2009. Colombia produced 13.2 billion cubic metres of natural gas in 2019, 0.3 percent of global supply (BP 2020).

With its oil reserves diminishing, it has been estimated Colombia can only support current levels of production for another six years (BP 2020; see also Smith 2018). This has led to concerns about the future of the industry and sparked heated political debates on the issue of fracking, still only in the exploratory stages in Colombia, with arguments relating to export revenues, energy independence, and the environment shaping the discussion. The Colombian Petroleum Association has estimated that about US$7 billion of investment would be needed each year to maintain current production levels (Murphy and Acosta 2018). With regard to coal, in 2018 Colombia had proven reserves of 4.4 billion tonnes, meaning current production levels could be sustained for 55 years (BP 2020). Capital investment in the Colombian gas sector has also decreased in recent years due to declining global prices, and its gas reserves of 100 billion cubic metres will only last for another eight years (BP 2020).

Despite its major hydrocarbon reserves, fossil fuels play a relatively minor role in electricity generation in Colombia, with natural gas and coal each accounting for about 10 percent of electricity generation in 2019. Oil plays only a minor role in electricity generation, though energy capacity additions since 2010 were mostly composed of hydropower, oil, and diesel (Climatescope 2020; IEA 2020). The Colombian power sector has a low-carbon footprint due to the high proportion of hydropower in its energy basket. In 2019, mainly large-scale dams provided 62 percent of the country's installed capacity and 71 percent of electricity generation (Climatescope 2020; see also IHA 2018). It is also continuing to build out its hydropower capacity, although the focus has recently been more on smaller-capacity projects (IHA 2018). Other renewable energy technologies, such as wind and solar PV, have so far only accounted for small shares of capacity and electricity generation, albeit increasing in recent years (IEA 2020; IRENA 2021a).

The electrification rate in Colombia is high at close to 100 percent (IEA 2020), and energy demand has risen steadily over the past decade. Further increases have been projected, despite significant progress in increasing energy efficiency among household and industrial users and in the power and transport sectors (GRICCE 2021). The latter is projected to remain the sector with the largest energy needs up to 2050 (Paez et al. 2017).

The high proportion of hydropower in Colombia's electricity mix has meant that it has largely been able to exploit its hydrocarbon reserves for export. In 2019, fossil fuel exports generated US$24.2 billion in revenue, and fuel exports accounted for about 55 percent of Colombian merchandise exports, although down from US$43.4 billion at the height of the oil price boom in 2013 (Chatham House 2021; World Bank 2021c; see Figure 5.1). Crude oil and coal exports accounted for the vast majority of these earnings, with Colombian natural gas mainly destined for domestic consumption.[1]

With Colombian oil reserves dwindling, investment in exploration low, and the future of Colombia's nascent fracking industry still uncertain, earnings from Colombia's to date most valuable export commodity may decrease over the coming decade. A key barrier

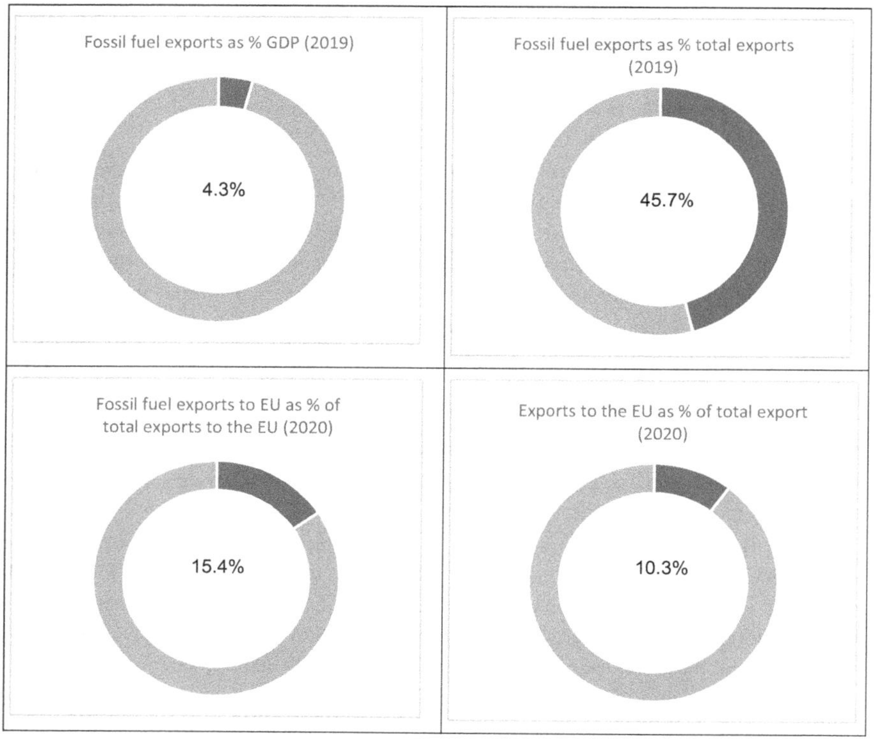

Figure 5.1 Status of fossil fuel exports of Colombia in 2019/2020.

Sources: Data retrieved from https://data.worldbank.org (fossil fuel exports as percentage of merchandise export; oil, gas, and coal as a percentage of GDP – accessed on 25 July 2021) and European Commission (2021a) (exports to EU – accessed on 25 July 2021).

to investment is that Colombian oil is relatively expensive to produce, with Colombia's largely state-owned oil company Ecopetrol and foreign field operators requiring the international oil price to stay above US$50 to remain profitable (Slav 2017). Oil industry investment and export revenues took a tumble when the oil price slumped from almost US$100 in 2014 to US$30 in 2016, and it remains to be seen if the Duque government's efforts to revive the sector will bear fruit (Acosta and Cobb 2019). Colombian crude oil exports generated US$14.6 billion in 2019, down from US$30.1 billion in 2013. If oil reserves were to decline, reduced exposure to such external price shocks could be a silver lining to falling oil rents. In 2019, about 45 percent of Colombia's oil exports went to the United States, with China and Panama being other significant export markets for Colombian crude oil (Chatham House 2021).

Coal has grown to become a key export commodity since 2000, and Colombia is the world's fourth largest exporter of thermal coal (IEA 2017a). In 2017, all but a tenth of the coal produced in Colombia was exported, generating a fifth of the country's foreign exchange earnings (Strambo and Atteridge 2018). However, future demand for Colombian coal exports is likely to rely on the development of new export markets. While in 2019 the EU was still Colombia's largest export market for coal, accounting for nearly 20 percent of its coal exports, EU coal imports decreased significantly between 2011 and 2019 (Chatham House 2021; Trademap 2021). The consumption of hard coal has halved in the EU since 1990, and demand looks set to fall further, as many EU member states move to reduce power sector emissions by phasing out coal-fired power plants, and some, such as Poland and Czechia, seek to support their domestic coal industries (Oei and Mendelevitch 2016). Demand for coal imports is also falling in what has traditionally been Colombia's other main export market, the United States (Trademap 2021). Although demand for coal is still increasing in Asian markets, Colombian coal is unlikely to be able to compete against coal from Indonesia and Australia, which can be extracted and transported at lower cost (Strambo and Atteridge 2018). Colombia is therefore likely to become increasingly reliant on just a few export markets. In 2017 and 2018, it increased its coal exports by significantly increasing its exports to Turkey and other Latin American countries, although coal exports in 2019 fell back to 2016 levels (Chatham House 2021; Trademap 2021).

Colombia has also faced significant budgetary pressures – due both to fluctuations in its key commodity markets and the major public spending commitments made by the government under the peace agreement with the FARC. Keeping the economy stable is therefore essential to honouring these commitments and Colombia's post-conflict development.

Future drops in oil rents, either due to falling reserves or oil market fluctuations, combined with the decline of the Colombian coal industry could subject the economy and public budgets to further pressures. Coal has contributed the largest part of the income generated by the extractives sector (Strambo and Atteridge 2018). However, economic growth and public budgets would be most vulnerable in the Departments of Cesar and La Guajira, which produce and export 90 percent of Colombian coal production. In 2015, coal accounted for about 40 percent of gross domestic product (GDP) and 30,000 direct jobs in these two departments. A further 100,000 jobs reportedly rely on the mines in the other departments. There has also been little focus on this issue in public debates or policy documents at national, departmental, and municipal levels, and as such the likely decline of the coal industry could lead to hardship in coal-producing regions (Strambo and Atteridge 2018).

Consecutive crises have accelerated the decline of fossil fuel industries in Colombia, in particular coal, and have led to policy adaptations. The fall in commodity prices, including coal and oil between 2014 and 2016, resulted in the contribution of fuel exports to GDP decreasing from 11 percent in 2013 to 6 percent in 2016.[2] As a knock-on effect on government budgets, the fiscal contribution of Colombia's extractives sector slumped from 19 percent of Colombia's national fiscal income in 2013 to just 5 percent in 2016 (EITI 2018). The subsequent stabilisation of the economy has been underpinned by important structural reforms including two sets of tax reforms in 2014 and 2016 (IMF 2018). The Covid-19 pandemic then led to a further decline of extractive industries and in particular coal. Colombia sled into a deep recession in 2020 with GDP contracting by 6.8 percent (World Bank 2021d). Coal production even plunged by 40 percent and may not fully recover given both domestic contestation and structurally declining international demand (Natural Resources Governance Institute 2021). While the economic recovery stimulus has a mixed environmental record, it has provided little support to a revival of the coal industry (Energy Policy Tracker 2021; Global Recovery Observatory 2021).

Other fragility and security risks

Still classed as a 'warning' area in the Fragile States Index, Colombia had become progressively more stable in the 2010s. There had been significant improvement in the strength of its democratic institutions and governance and in its moves towards ending the longstanding internal conflicts with insurgent groups, most notably the peace agreement with the FARC in 2016 (Fund for Peace 2021). However, the peace agreement has become increasingly fragile, its implementation has degraded under the administration of President Duque, and corruption remains a barrier to effective governance. As a result, also Colombia's assessment in the Fragile States Index has deteriorated in 2020–2021 (Fund for Peace 2021). Climate change may also intensify existing fragility risks in rural areas and impair hydropower electricity generation (see Figures 5.2 and 5.3).

Complex internal conflicts have for long constituted the major national challenge for Colombia. Achieving long-term peace with the FARC, and other insurgent groups, will depend on the successful implementation of the 2016 peace agreement, which covered six main areas – comprehensive rural development, illegal crop eradication, the FARC's political participation in Colombian politics, transitional justice and reparations for victims, and the demobilisation, disarmament, and reintegration of ex-combatants (Felter and Renwick 2017). However, the implementation of the peace process has significantly deteriorated since President Duque took office in 2018 and has suffered further setbacks under the Covid-19 pandemic.

Colombia's long and complex internal conflict has been shaped by a number of political and socioeconomic drivers. A major root cause of the conflict was the issue of land tenure and land concentration. Extremely unequal land ownership had been a longstanding cause of political tensions since colonial times, and FARC and ELN insurgents gained legitimacy and recruits by opposing the privatisation of natural resources and striving for the redistribution of land to small peasants and the abolition of large landholdings (Bilotta 2017). The peace agreement set out a series of bold initiatives to comprehensively reform and

Fragility (2021)	Sustainable	Stable	Warning	ALERT
Human development (2019)	Very high	**HIGH**	Medium	Low
Strength of governance (2019)	Very high	High	**MEDIUM**	Low
Climate change vulnerability (2019)	Low	Medium	**HIGH**	Very high
Sustainable energy development (2019)	Very high	**HIGH**	Medium	Low

Figure 5.2 Governance indicators (Colombia).

Sources: Fund for Peace (2021), ND-GAIN (2020), UNDP (2020), World Bank (2021e, 2021g).

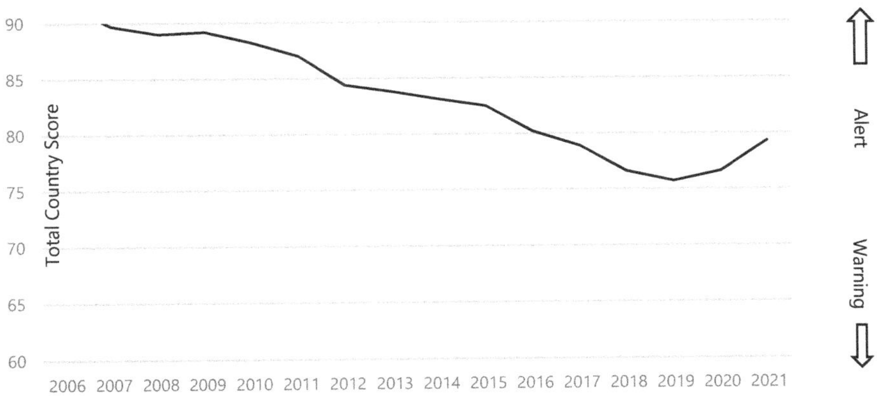

Figure 5.3 Colombia's performance in the Fragile States Index 2006–2021.
Source: Fund for Peace (2021).
Note: Higher index numbers indicate a higher level of state fragility.

revitalise Colombia's rural areas, which experts estimated could cost between US$80 and US$90 billion over the subsequent decade (Felter and Renwick 2017). Land is also central to reparations under the peace agreement, with an estimated 8 million hectares of land – 14 percent of Colombia's territory – illegally acquired either by dispossessing or forcibly displacing people over the course of the conflict (Amnesty 2014; OXFAM 2016). Coca (i.e., the raw material for producing cocaine) and cocaine production was also a key driver of the conflict due to its role in financing the activities of rebel groups, and it continues to present complex challenges for the peace process. Providing economic alternatives to the drugs trade is of crucial importance, but it is fraught with difficulty as other insurgent groups and cartels compete to take control of coca-growing areas previously controlled by the FARC and coca farmers lack alternatives (Bristow 2018).

The conflict and the peace process continue to shape the country's politics. President Duque was elected in June 2018 on a platform that included harsh criticism of the peace agreement. Since taking office, he has been openly hostile to the peace agreement. His administration has generally been slow in implementing the agreement, and the remaining challenges are daunting (International Crisis Group 2019a; Fund for Peace 2020; Reith 2021). The Covid-19 pandemic has further accentuated these challenges, as it 'laid bare the shortcomings in implementing the FARC peace agreement' and strengthened other armed groups (International Crisis Group 2020).

The sustainable use of Colombia's immense natural resources also has a crucial role to play in Colombia's post-conflict development (UNEP 2017), and the peace agreement set important precedents with specific actions regarding environmental sustainability, natural resource management, and progressive measures on gender. However, FARC's demobilisation and the end to violence also opened up a vacuum in terms of environmental governance, and, in 2016, deforestation increased by 44 percent as smallholder farmers and industry rushed to take control of the jungle and the Amazon, to log it and convert it for uses such as cattle ranching and goldmining (Reardon 2018).

The Venezuelan crisis has added a further major challenge for Colombia in the late 2010s. By the end of 2020, Colombia was hosting more than 1.7 million of the 4.5 million Venezuelans who had fled their country since the economic and political situation severely deteriorated in 2017 (UNHCR 2020). These refugees added to the approximately 8 million Colombians who have been displaced over the decades due to domestic conflict (UNHCR 2020, 2021). The government has provided support in the form of temporary shelter along the border, as well as temporary residency, and access to healthcare and education. However, it faces growing challenges and rapidly increasing costs to stem the emerging humanitarian crisis. International support only covered 24 percent of the estimated €280 million required to provide for the Venezuelan refugees' basic needs in 2019. Consequently, there is growing pressure on public budgets, and the government was 'contemplating relaxing deficit targets so that it can spend an extra €800 million on meeting the needs of Venezuelan refugees and reallocating money that should be spent on other priorities, including implementation of the 2016 peace deal' (International Crisis Group 2019b). The Covid-19 pandemic and resulting crisis have further increased the precariousness of the situation of the Venezuelan refugees.[3]

Despite the aforementioned conflicts, over the past decades Colombia has developed relatively strong democratic institutions. There have been free and fair elections in Colombia for nearly 50 years, and, apart from a few notable exceptions, such as vote-buying and intimidation, there are no major concerns regarding voting rights, party competition, and election campaigns (Bertelsmann Stiftung 2020). Looking at the World Bank's Worldwide Governance Indicators, over the past decade there have also been gradual improvements in regulatory quality, and voice and accountability in Colombia, although this index shows that overall few gains have been made in enhancing government effectiveness, the rule of law, and control of corruption (World Bank 2021e). Though still categorising Colombia as 'partly free', the Freedom House Index 2021 states that 'public institutions have demonstrated the capacity to check executive power and enforce the rule of law, and violence declined as the government and the country's main left-wing guerrilla group moved toward a peace accord signed in 2016'. However, the report also notes that Colombia still faces major challenges with regard to the consolidation of peace and the guarantee of political rights and civil liberties, especially on the countryside (Freedom House 2021; see also Figures 5.2 and 5.3).

Alongside poverty and inequality, corruption constitutes a further central challenge for Colombia (Bertelsmann Stiftung 2020). Each year, corruption is estimated to cost the Colombian economy US$17 billion – equivalent to 5.3 percent of GDP (Grattan 2018). As such, Colombia scored 39 out of a potential 100 on Transparency International's 2020 Corruption Perceptions Index, placing it 92nd out of 180 assessed states (Transparency International 2021). One driver of corruption may be the process of decentralisation in Colombia. Beginning in the 1980s and consolidated in the 1991 constitution, this process resulted in greater efficiencies and social spending, but it also created openings for organised crime groups. 'Following the 2015 local elections, nearly one in three governors' offices and nearly one in seven mayors' offices came under investigation for various suspected crimes, including ties to paramilitaries, drug traffickers and other criminal groups' (Yagoub 2016).

Colombia is also highly vulnerable to the slow and sudden impacts of climate change, and, given its varied geography, the country will have to cope with a variety of adaptation challenges (see also Figure 5.2). Projections show that rainfall may decrease by nearly a third in some areas, with the Andean region shifting from a semi-humid to a semi-arid climate (OECD 2014). Sea-level rise will impact coastal and marine ecosystems and fisheries as well as Colombia's coastal infrastructure and cities (US AID 2017). Extreme weather events and disasters such as landslides also represent major climate change threats. Like the rest of Latin America, Colombia is experiencing more frequent and severe La Niña and El Niño phenomena, which are characterised by intense periods of drought followed by heavy rain and flooding (FAO 2018). It has been estimated that 85 percent of the Colombian population and 87 percent of the country's GDP are 'at risk' from natural disasters (OECD 2014). The 2010–2011 La Niña led to estimated economic losses of approximately US$6 billion (NDC Partnership 2017).

Higher temperatures and increased water scarcity are of particular concern in the Andean region, home to 75 percent of the population. There, the run-off from the mountains is the main water source for domestic and industrial users, farming and irrigation, and electricity generation (OECD 2014). Indeed, 80 percent of Colombia's GDP is generated in the basin of the two largest rivers running through the Colombian Andes, the Magdalena, and the Cauca. The large-scale dams in the watershed hold 84 percent of the Colombia's hydroelectric power (Baptiste et al. 2017). Their vulnerability to climate change became evident during the 1992 and 2015–2016 El Niños. Droughts combined with forest loss reduced water levels to record lows and increased sedimentation. These conditions triggered major electricity crises and a sharp increase in the use of thermal plants (Semana 2015).

In the agricultural sector, farmers may have to move high-value agricultural crops such as coffee to higher altitudes to achieve the same yields, and other important crops, such as tropical fruit, cocoa, and bananas are also at risk (US AID 2017). The impact of climate change on livelihoods in these areas will be shaped by the socioeconomic struggles that influence activities like illegal land acquisition and conversion of forests to agricultural farms or cattle breeding grounds and pasture lands. For example, communities' vulnerability to extreme weather events is 'strongly influenced by deforestation, slash-and-burn agriculture, artificial drainage of wetlands, changes of natural river courses and the building of human settlements in areas at risk of floods or landslides' (OECD 2014).

Past and present efforts to decarbonise

Colombia has been one of the most active developing countries in the international climate negotiations (Bustos 2017). It ratified the UN Framework Convention on Climate Change (UNFCCC) in 1994 as a non-Annex I party to the Convention. Colombia ratified the Kyoto Protocol in 2005. It has pushed for greater ambition under the UNFCCC as a leading member of the AILAC group of progressive Latin American countries, the Cartagena Dialogue, the Climate Vulnerable Forum, and as part of the High Ambition Coalition. Its congress unanimously ratified the Paris Agreement in July 2017 (WWF 2017).

Colombia submitted its intended nationally determined contribution (NDC) to the Paris Agreement in September 2015, committing to unconditionally reducing its emissions by 20 percent compared to the projected business-as-usual scenario by 2030. Furthermore, Colombia would increase its commitment to a 30 percent emission reduction on the 2030 business-as-usual scenario, conditional on international support. Mitigation, adaptation, and means of implementation are all addressed in Colombia's NDC, with adaptation a priority due to the country's high vulnerability. Mitigation is most important in the energy sector and the agriculture, forestry, and land use sectors, but eight Sectoral Mitigation Action Plans have been developed to reduce emissions across all sectors specified by the Intergovernmental Panel on Climate Change (IPCC) (NDC Partnership 2017). Emissions projections are strongly influenced by emissions from land use, land use change, and forestry (LULUCF), which are subject to considerable uncertainties, but, in 2019, Colombia was considered on track to achieve its unconditional NDC target (Kuramochi et al. 2019).

In December 2020, Colombia submitted an updated NDC. It thereby increased its overall emission reduction target to 51 percent below business as usual by 2030. At the same time, the strategy for achieving the 2030 target relies heavily on action in the LULUCF sector, whereas action in other key sectors (such as energy supply and transport) receives insufficient attention. The government of Colombia has also established a long-term target of achieving carbon neutrality by 2050 (the implementation of which will require further action) (CAT 2021).

In 2018, Colombia's greenhouse gas (GHG) emissions amounted to 267.95 Mt of CO_2 equivalents, accounting for 0.55 percent of global emissions (WRI 2021). In that year, energy, LULUCF, and agriculture were the highest emitting sectors, accounting for approximately 34 percent, 31 percent, and 23.5 percent of the total emissions, respectively (WRI 2021; see Figure 5.4).

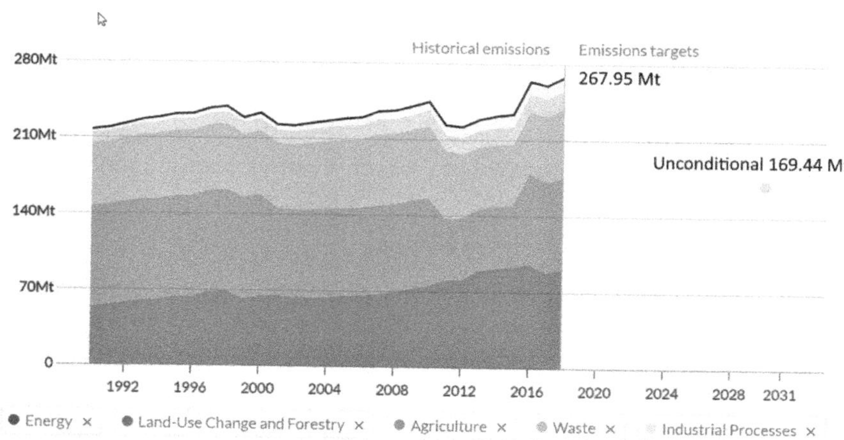

Figure 5.4 Colombia's greenhouse gas emissions and emission targets (1990–2018).
Source: WRI (2021).

Colombia has made good progress in developing robust institutions, policies, and laws to mainstream climate action, environmental sustainability, and green growth strategies into its economic development. Colombia has developed a robust institutional, legal, and policy framework for implementing its NDC (NDC Partnership 2017). In 2016, the National Economic and Social Policy Council played a leading role in establishing the National Climate Change System (Sistema Nacional de Cambio Climático, SISCLIMA), which aims to provide a comprehensive institutional framework for the implementation of the Colombian government's main climate strategies, including a Climate Change National Adaptation Plan, a Colombian Low Carbon Development Strategy, a National REDD+ Strategy, and a Strategy for Fiscal Protection Against Natural Disasters (GRICCE 2021). SISCLIMA coordinates Colombian climate action at local, departmental, national, and international levels (GRICCE 2016; LSE 2019) and brings ministries together with the private sector and non-state actors (NDC Cluster 2018). Since passing the Guidelines for Climate Change Policy in 2002, Colombia's government has adopted successive legislation and laws to strengthen the country's climate change mitigation and adaptation action, with a significant focus on promoting market mechanisms. Key pieces of recent legislation include the National Policy on Climate Change and the introduction of a carbon tax in 2017 and the Green Growth Policy introduced in 2018. The country's first climate change law was passed in July 2018.

The Colombian government has conducted analysis of the economic benefits of climate action for Colombia's long-term growth (DNP 2015) and taken steps towards mainstreaming climate action and environmental sustainability within Colombia's overarching economic development plans. The last three National Development Plans have prioritised the sustainable use of Colombia's national resources. Introduced by the Duque administration in autumn 2018, the latest National Development Plan for 2018–2022 acknowledges climate change as one of the key 'constraints on the country's progress' and comprises a 'pact for sustainability' that outlines, for example, national priorities for climate change mitigation and the circular economy (DNP 2018). While the plan's exact implementation remains to be seen, the pact's provisions have been criticised as 'not precise and lacking ambition' – for example, due to the absence of strategy to halt deforestation. The plan also includes a 'Pact for mining-energy resources for sustainable growth and the expansion of opportunities', which foresees an increase in coal extraction, hydrocarbon production, and oil exploration (Peñaloza 2019).

Colombia challenges the idea that middle-income countries dependent on fossil fuel exports are not able to play a constructive role in international climate negotiations (Bustos 2017). This is the result of the dynamics between the different institutions and pressure groups active in domestic climate politics. The 2018 presidential elections provided a snapshot of these different coalitions of interest, with the future of the country's fossil fuel industries a key issue distinguishing the two candidates. While leftist candidate Gustavo Petro pledged to phase out coal and oil exports for both economic and environmental reasons, President Duque supported the continued development of the country's oil, coal, and gas sectors (Casey and Abad 2018). In July 2018, he appointed former Executive Vice President of state oil company Ecopetrol[4] Maria Fernanda Suarez Londoño as Minister of Mines and Energy, and he has encouraged investment in exploration by offering tax relief to the industry. His administration continues to support initiatives to explore whether fracking can be economically and environmentally viable (Acosta and Taj 2019) but has also taken action to realise the previous Santos administration's plans for renewable energy auctions. Nonetheless, the changing political economy of coal mining and the huge investment needed to maintain current levels of oil production have eroded the strong support and institutional framework that has supported fossil fuel extraction and mining activities in Colombia for decades (Strambo 2017).

Among the public, there is widespread acceptance that climate change is happening and over two-thirds believe that climate change is the result of human activities. A focus group study suggested that Colombians are strongly in favour of measures to mitigate and adapt to the effects of climate change and are most likely to associate climate change with the more frequent and severe flooding and droughts that had occurred in recent years. They are concerned about the negative impacts on the agricultural activities and quality of life (Martínez and Alfonso 2018).

Trends and potential

Colombia's future prosperity will rest on three pillars: the peace process, economic stability, and environmental sustainability. Colombia's updated NDC has a major role to play in this, and its implementation is being financed in part by climate finance, a carbon tax, and, in future, an emissions trading system. Provided the peace process advances, the economy also looks set to reap a 'peace dividend' – first and foremost in the agricultural sector. The 'orange economy', comprising the innovation and creative industries, remains a major focus, with its development reliant on high educational standards.

Financing low-carbon development

Colombia has been successful in attracting climate finance from a range of streams, including international donors and domestic sources of public and private climate finance. The financial committee of SISCLIMA has been meeting regularly since 2013 and has developed a robust monitoring, reporting, and verification system for the country's climate change project financing. The process has been led by the National Planning Department, with the assistance of the Ministry of Foreign Affairs, and other institutions have gradually been invited to become members (GRICCE 2021). Various international donors have provided financial support ranging from bilateral instruments such as Germany's International Climate Initiative (IKI 2018) to the Green Climate Fund (Duque et al. 2017; Green Climate Fund 2018) and multilateral initiatives such as the US$100 million that Germany, Norway, and the United Kingdom pledged to support Colombia's efforts to reduce deforestation in the Amazon under the UNFCCC's Reducing Emissions from Deforestation and Forest Degradation (REDD+) scheme (GGGI 2015).

Carbon market mechanisms

At domestic level, Colombia is using green taxes and market mechanisms to support the country's sustainable development and the ongoing peace process, as well as to compensate for a potential decline in oil tax revenues. The major tax reform passed by the

Colombian Congress in 2016 included a new carbon tax to be levied on all sales and imports of fuels, including all petroleum derivatives, but excluding coal and consumption of natural gas in electricity generation. As such, it covers half of Colombia's fossil fuel emissions and 16 percent of its emissions overall. The tax entered into force on 1 January 2017 at US\$5/t CO_2e – below average when compared with other carbon taxes around the world – but is set to gradually increase to US\$10/t CO_2e. The tax is estimated to generate around US\$200 million in revenue each year, which is channelled into the 'Colombia in Peace Fund' administered by the Ministry of Environment and Sustainable Development. Around 70 percent of the funds are directed towards peacebuilding initiatives meeting certain sustainability criteria and a quarter of the funds towards environmental activities related to coastal erosion, reduction of deforestation, conservation of water sources, conservation of strategic ecosystems, and climate change. The remaining 5 percent of funds are allocated to a national heritage fund for conservation of biodiversity (IETA 2018; Monge 2018).

Offset schemes and emission certificates are also being developed. Since June 2017, companies and other types of organisations have had the option of offsetting their fuel consumption and obligations under the carbon tax by buying certificates generated, for example, by UN REDD+[5] projects (Conservation International 2018). This has in turn led to the development of more projects designed to mitigate carbon emissions. In the first six months, these offsets compensated around 2 Mt CO_2, equivalent to around 5 percent of the expected tax collection (Alarcon-Diaz and Lubowski 2018).

Furthermore, Colombia has taken initial steps towards establishing an emissions trading scheme in the medium term. An assessment looking at the potential economic advantages of different design options has been initiated, and Congress laid the groundwork for creating a system of carbon credits and allowances in the Climate Change Law passed in July 2018. First steps towards creating a platform to register and trade verified emissions have been taken (ICAP 2021).

Potential in key low-carbon sectors

In July 2018, the Santos administration approved a Long-Term Green Growth Strategy for Colombia. Recognising that Colombia's current economic development model is depleting its resource base and generating high costs for the environment and society – estimated at more than 16.6 billion pesos, equivalent to 2.08 percent of GDP in 2015 (DNP 2018) – the strategy aimed to increase the economy's productivity and competitiveness up to 2030, while also achieving the Sustainable Development Goals (SDGs) and the emission reductions outlined in Colombia's NDC. Developed in collaboration with international partners[6] and Colombian stakeholders, the strategy specifies key sectors and cross-cutting priorities for the economy, which include developing agribusiness and forestry; expanding and diversifying the use of clean energies; increasing efficiency in the use of water, soil, and energy; promoting the circular economy; and encouraging science, technology, innovation, and green jobs (GGGI 2018). The strategy launched a wide-ranging implementation process comprising 150 actions to be carried out by over 25 ministries and agencies. However, there has been criticism that its objectives have not sufficiently been incorporated into the 2018–2022 National Development Plan (Peñaloza 2019).

Although the importance of the agricultural sector in the economy has declined in recent decades, the peace process has opened up opportunities for future investment and growth. Large areas of land previously controlled by the FARC became accessible and, combined with Colombia's favourable climate and water resources, could represent a significant peace dividend. The government's Green Growth Strategy estimated that the 'bioeconomy' sector could grow 2.5 percent annually, with the number of planted hectares expected to increase by 44 percent between 2015 and 2030 (Government of Colombia 2018). The National Association of Entrepreneurs of Colombia was even more optimistic, predicting an annual 8–10 percent increase in agricultural production, with major increases in crops such as African palm, cocoa, and mangos (Oxford Business Group 2018). Much of this increase in production could be exported, and in 2019 agricultural products were already worth US$7.2 billion, making up 21 percent of Colombian commodities exports (Chatham House 2021).

The rural reforms launched as part of the peace process will be crucial in determining both the development of Colombia's agricultural sector and the stability of the peacebuilding process. Agriculture and cattle ranching are important livelihoods in the rural areas affected by Colombia's internal conflicts. The peace agreement aimed to simultaneously increase the cultivation of licit crops and address longstanding grievances about land rights and land concentration by opening up previously uncultivated land to agriculture, including smallholder farmers and creating a land fund with 3 million hectares to formalise small and medium landholdings. It also aimed to strengthen the property rights and economic independence of rural women (OXFAM 2016; Oxford Business Group 2018). However, implementation of the reforms has slowed since President Duque took office. 'According to the Kroc Institute, which tracks peace agreement progress, 51 percent of the initiatives in the rural reform chapter have made such little progress that it is unclear they will ever be fully implemented and a further 38 percent have made no progress at all' (International Crisis Group 2019b).

As acknowledged in its (updated) NDC, measures to adapt to and mitigate the impacts of climate change can also be designed to play an important role in strengthening the peace process and the agricultural sector in Colombia. For example, promoting the cultivation of crops, such as cocoa, in already deforested areas previously used for coca production is preferable to clearing forest to make way for cattle ranching (CIAT 2018). Although many farmers in the Amazon are turning to the latter as an alternative to growing coca, it is unsustainable as ever greater areas of land are needed due to the unsuitable soil (Palmer 2017). Introducing climate-smart agricultural techniques will also be important in the main producing regions in Colombia's mountain and coastal areas (US AID 2017). For example, the land reforms agreed during the peace process included a plan to ensure 'democratic and environmentally sustainable access to water' for 'family-run and community-based' farming (Government of Colombia 2016).

As to renewable energy, domestic energy production in Colombia already has a very low carbon footprint due to the large proportion of hydropower in its electricity mix. However, the National Energy Plan for 2050 aims to diversify the country's renewable energy mix beyond hydropower to meet growing demand and ensure a reliable, low-emission electricity supply, even during the droughts in El Niño cycles. There is significant potential for the inclusion of wind power plants, photovoltaic solar generation, geothermal energy, and generation from biomass in the country's energy basket, and the plan's most feasible scenario estimates these other 'unconventional' sources could account for 6 percent of the electricity mix by 2028, equivalent to an installed capacity of 1,207 MW (IEA 2017b). The Duque administration drew up contracting projects for up to 1,500 MW of installed capacity, prioritising such non-conventional sources of renewable energy in its first months in office (Ministry of Mines 2018). It also planned to allocate 1.18 million MWh per year via auctions. The first large-scale auction for unconventional renewable energy capacity held in February 2019 failed to award any contracts since the auction's competition criteria were not met. However, a second auction held in October 2019 successfully awarded contracts for about 1.3 GW of new wind and solar photovoltaic capacity, expected to start operating by 2022 (Bellini 2019; IRENA 2021b). In addition, Colombia's economic recovery plan from the Covid-19 crisis foresees significant additional spending on renewable energy and clean growth (Energy Policy Tracker 2021; Global Recovery Observatory 2021).

The orange economy

During his presidential campaign, Duque frequently highlighted the importance of developing the 'orange economy' – entrepreneurship, innovative technologies, and the creative sector, including music, film, and fashion design – as new exporting industries and engines of the Colombian economy. Before entering politics, he co-authored an Inter-American Development Bank publication entitled 'The Orange Economy – An Infinite Opportunity', which highlighted that Latin America had yet to harness the economic potential of its creative industries, which had seen exponential growth in other parts of the world. It also made the argument that the creative industries were more profitable than extractives – stating that 'by 2012 activities at the core of the orange economy would be contributing US$2.2 trillion dollars per year to the world economy, ... equivalent to 230 percent of the actual value of the oil exports of all the OPEC members for the same year' (Duque and Buitrago 2013). Once in office as a senator, Duque also wrote the 'Orange Law', passed in May 2017, to promote, develop, and protect Colombia's creative industries. In 2017, the industries associated with the orange economy accounted for 3.4 percent of Colombia's GDP, more than mining or coffee (Dempsey 2020), and the National Development Plan 2018–2022 targets growth to 6 percent by 2022 (DNP 2018).

Education and skills development

Both access to education and the quality of teaching in Colombia have improved significantly over the past two decades, and, according to the World Bank's Human Capital Index, the current level of education and healthcare would allow the average child born in Colombia today to achieve around 60 percent of their potential, somewhat above the worldwide and Latin American average (World Bank 2020a). Successive reforms have boosted enrolment rates in primary, secondary, and higher education, and increased standards in the teaching profession, for example, by raising entry requirements or channelling investment into skills development. New governance structures and funding arrangements have also made the education system more efficient (OECD 2016). In 2018, the literacy rate stood at around 95 percent, and the average number of years spent in primary to tertiary education was 14 years for males and 15 years for females (CIA 2021).

Ensuring a lasting peace and sustainable development in Colombia will depend on many factors, but it has been argued that 'none will be more important to the country's future than its ability to build a strong and inclusive education system' (OECD 2016). The 2018–2022 National Development Plan lists education as one of the main pillars of Colombia's future prosperity (DNP 2018). However, to achieve the government's goal of becoming the 'most educated' country in Latin America by 2025, further investment and changes are needed. In 2018, expenditure on education amounted to 4.5 percent of the Colombian GDP, slightly below the average of the member countries of the Organisation for Economic Co-operation and Development (OECD) (5 percent in 2017) and the EU (4.7 percent in 2017) (World Bank 2021f). The strong inequalities between socioeconomic groups and regions in Colombia are also more than evident in its educational system. Depending on a child's location and background, the number of years a child spends in formal education can range from 6 to 12 years (OECD 2016). These disparities also affect the enrolment rates in tertiary education across different socioeconomic groups. While 53 percent of students from wealthy families enrol at universities, only 9 percent of students from poor families do so (OECD 2016).

Low skills and lack of access to training present major barriers to those seeking decent jobs in the labour market. Many low-skilled workers end up in Colombia's large informal labour market, which is estimated to employ between 50 and 60 percent of the workforce (Bertelsmann Stiftung 2020). Unemployment in the pre-Covid-19 period was slightly above 10 percent before surging to more than 20 percent at the height of the Covid-19 crisis in 2020 and retreating to around 15 percent since (Trading Economics 2021). The distribution of employment has not changed significantly over the past decade, and, in 2020, 63.5 percent were employed in the services sector, 20 percent in industry, and 16.5 percent in agriculture (Statista 2021). Besides policies to level out the enrolment rates between regions and socioeconomic groups, the OECD has argued that improving the quality of technical and technological education should be a priority (OECD 2016).

Cooperation with the EU

Cooperation between Colombia and the EU is well developed, with Colombia being one of the EU's most important partners in Latin America. After the United States and China, the EU is Colombia's most important trade partner and has also been a strong supporter of Colombia's peace process and post-conflict development. Fossil fuels represent a major pillar of EU-Colombia trade, and the reduction in European coal demand is leading the Colombian coal sector to seek new export markets.

Institutional framework: multiparty free trade agreement

As Colombia has developed into an upper-middle-income country, the focus of bilateral relations has shifted from development cooperation towards political dialogue and trade. Trade between Colombia and the EU is governed by the multiparty free trade agreement, which was formalised with Colombia and Peru in 2013, and Ecuador in 2017 (Grieger and Harte 2018). Overall, it is considered to have had a 'stabilising' and 'clear positive impact' on trade between the EU and the three Andean countries (EPRS 2018). The EU is Colombia's third biggest trading partner after the United States and China, and in 2019 bilateral trade totalled €10.5 billion (€8.3 billion in 2020), with the EU importing €4.3 billion (€3.7 billion in 2020) worth of goods from Colombia (European Commission 2021a, 2021b). Since the free trade agreement came into force in 2013, Colombia has diversified its exports to the EU (European Commission 2018). Since 2019, agricultural products have substituted fossil fuels as the largest share of Colombian exports to the EU. According to the European Commission, fuels and mining products accounted for 16.2 percent of imports from Colombia in 2020 (when they took a particular hit due to the Covid-19 pandemic), compared to almost 50 percent in 2017 (European Commission 2021a, 2021b; see also Figure 5.1). Coal accounted for more than 60 percent of fossil fuel exports (Chatham House 2021). Agriculture constituted 57 percent of total exports to the EU in 2020, up from 42 percent in 2017 (European Commission 2021a). This has had a positive effect on the implementation of the peace agreement by improving regional employment opportunities (European Commission 2018). The agreement has also allowed Colombian small and medium-sized enterprises (SMEs) to increase their exports to the EU (European Commission 2018). Colombian imports largely comprise manufactured goods, equipment, pharmaceuticals, and chemicals (Garcia 2016).

European investments have mainly been in the services sector and directed towards companies, software, IT, and finance (ProColombia 2018). Colombia's membership of the OECD is likely to yield further opportunities for cooperation on modernising the economy, for example, in reforming its education system and labour market, and the reforms that accompanied OECD membership, such as those to reduce corruption, have made the country a more attractive destination for European foreign investment. In 2020, Colombia achieved the third highest score in the Latin America and Caribbean region on the World Bank's Ease of Doing Business Index (World Bank 2020b).

The free trade agreement is an example of a 'new generation' of free trade agreements, in which the EU has sought to promote sustainable development in its external trade relations, and includes clauses on human and labour rights, environmental protection, and good governance. At the time it was negotiated, the agreement was welcomed by business associations but criticised by civil society groups for its lack of detail regarding the enforcement of the sustainable development clauses.[7] Five years later, an assessment on behalf of the European Parliament concluded that 'serious concerns' remained about the agreement's ability to ensure that 'environmental standards will be met and that those standards will be relaxed to facilitate investment in extractive industries'. The report recommended closer cooperation in international and multilateral fora relevant to mining, energy, and hydrocarbon sectors, as well as the monitoring of good practices among European enterprises in Colombia, as potential entry points for promoting sustainable development (EPRS 2018).

Cooperation on climate action and clean energy

Colombia and the EU are strong partners with 'very like-minded views' in international processes related to climate change, and, under the UNFCCC, Colombia and the AILAC group have coordinated efforts to increase ambition in the negotiations via the Cartagena Dialogue and the High Ambition Coalition. This also applies to broader processes relating to Agenda 2030 and sustainable development, the sustainable use of natural resources, and the protection of biodiversity (EEAS 2018a). For example, cooperation via the International Urban Cooperation programme has linked up European and Colombian cities, so that they can share knowledge and experience for meeting the goals of the New Urban Agenda, the Paris Agreement, and the SDGs (European Commission 2016). The diversification of Colombia's renewable energy mix could provide an entry point for further cooperation, as it has remained relatively underdeveloped to date.

Support for the peace process

A major focus of EU-Colombia relations to date has been supporting Colombia on the road to peace. Since 2002, the EU and its member states have provided €1.5 billion to support peacebuilding in Colombia, €550 million of which came from the EU budget. This has been used to promote reconciliation and address the root causes of the conflict by investing in economic and regional development, human rights, and democratic governance at local level. The EU also provided support throughout the peace process with the FARC but was particularly active during the final six months of the negotiations when the Colombian government reached out for international partners to support Colombia's post-conflict development. At this stage, former High Representative Federica Mogherini appointed a special envoy to coordinate the EU response and support, allowing for a more rapid, organised, and strategic cooperation at a critical time (Chaparro 2018). The EU has since reaffirmed its support for Colombia's post-conflict development with an overall support package totalling almost €600 million. This includes €95 million for the EU Trust Fund for Colombia agreed in December 2016 to support 'comprehensive rural development', reinforce 'state presence and restore the social fabric in areas affected by the conflict', and reintegrate ex-combatants (EEAS 2018a). Building on its long history of involvement in the process of conflict resolution and peacebuilding, the EU continues to hold an important potential to contribute to keeping the precarious peace process alive and advancing it (Ioannides 2019; International Crisis Group 2020).

Broader cooperation agenda

The EU and Colombia are engaged in a number of political and policy consultations at bilateral, regional, and global levels. A bilateral high-level political dialogue in July 2019 saw the two sides affirm their commitment to continued cooperation on a range of areas of mutual interest, including human rights, security and defence matters, protection of the environment, education, science and innovation, and cooperation in the fight against illicit drugs. They also agreed to explore new areas of cooperation, including the creative industries (the 'orange economy') and the circular economy, and new mechanisms for 'lifting bilateral relations to a new level' in future (EEAS 2019). The latest round of bilateral high-level political dialogue in December 2020 focused on the Covid-19 pandemic and the need for inclusive and sustainable green recovery. Both parties expressed satisfaction with the progress made towards the finalisation of a Memorandum of Understanding on an agenda of enhanced political and sectoral dialogue and cooperation for the next decade in which an environmental and a digital agenda are supposed to feature strongly (EEAS 2020).

In the area of security and defence, the Framework Participation Agreement that the EU and Colombia concluded in August 2014 has also made way for greater future cooperation between both partners in missions undertaken within the framework of the Common Security and Defence Policy. Colombia has also asked for further EU assistance to help respond to the Venezuelan crisis.

EU-Colombia relations are also shaped by various dialogues promoting cooperation within the Community of Latin American and Caribbean States. These include the Joint Initiative on Research and Innovation, the Erasmus+ programme for cooperation in higher education, a Structured Dialogue on Migration, and a Coordination and Cooperation Mechanism on Drugs (EEAS 2018b). In 2015, Colombia's Ministry of External Relations and the European Commission's Directorate-General for Regional and Urban Policy also established regional policy dialogues to support the strengthening of economic, social, and territorial cohesion – an important aspect of Colombia's post-conflict development (European Commission 2015).

Conclusions

After decades of internal conflict, Colombia made an important step towards peace in the 2016 peace agreement with the FARC. However, the process of building lasting peace is in its early stages, has become increasingly fragile, and has come under additional strain as a result of the Covid-19 pandemic. Continued political commitment is needed, and economic stability will also be crucial, given the high cost of fulfilling the terms of the peace agreement. Major public spending is required to enact, for example, the rural reforms that are fundamental to the peace deal and Colombia's post-conflict development, as well as to respond to other security challenges, such as narcotics production and trafficking, continued activities of other insurgent groups, and large influx of Venezuelan refugees. Oil and coal remain important to the economy, accounting for a substantial proportion of Colombian exports. It remains to be seen how prices recover from the pandemic but for the moment it appears that the unfavourable economic conditions have tempered the Duque administration's focus on expanding fossil fuel production.

Global decarbonisation processes therefore have the potential to reinforce economic and political challenges to stability in Colombia. With decarbonisation precipitating falls in demand for coal imports in its traditional export markets, including the EU, and Colombian coal lacking competitiveness in other major coal markets, Colombian coal exports will likely rely on a shrinking circle of countries that still favour coal power. With no coal exit strategies or policies in place as of 2021, any significant short-term decline in coal exports could harm the economy and present significant economic and development challenges in coal mining regions. Furthermore, although Colombia proved more resilient than other countries to the sharp drop in oil prices between 2014 and 2016, future market fluctuations combined with a potential decline of its coal exports could place pressure on public budgets, with implications for political stability. Decarbonisation therefore presents the challenge of diversifying away from fossil fuel extraction and developing a strong low-carbon economy, on top of the complex task of implementing the fundamental reforms required for the success of the peace process. The fallout of the Covid-19 pandemic has given a strong foretaste of the size of the challenge since 2020.

The EU and Colombia have well-established relations and are cooperating in many areas that can support both decarbonisation and Colombia's post-conflict development. The EU has offered solid support to Colombia throughout the peace process and has launched initiatives that can play a role in supporting long-term peace, particularly in the crucial area of rural development. Trade relations between Colombia and EU member states have already advanced beyond coal, as trade in agricultural products has increased under the multiparty free trade agreement, also supporting rural development. Although the sustainable development clauses of the free trade agreement have played a limited role in strengthening environmental governance to date, cooperation in other forums could begin to fill this gap. The EU and Colombia have also cooperated constructively in climate policy. With Colombia an example of an oil-producing state playing an active and constructive role in the international climate negotiations, it has also cooperated with the EU to advance decarbonisation by increasing ambition in the UNFCCC.

On the basis of our analysis, we can identify rural development, climate action, and economic diversification as important entry points for EU-Colombia cooperation to support both decarbonisation and post-conflict development processes. Accordingly, the following areas may constitute key priorities for European foreign policy towards Colombia:

- **Political and diplomatic support for peace and post-conflict development**: The EU has assured a reliable, coordinated support for the peace process with the FARC using a range of diplomatic instruments, including appointment of a special envoy. The EU can continue to make use of its diplomatic toolbox to develop sound European responses to support lasting peace, as well as responses to other security challenges, such as the Venezuelan crisis.
- **Economic diversification**: To support the transition towards a low-carbon economy in Colombia, the EU can support initiatives and investment advancing and strengthening commercial ties in the alternative sectors proposed by the Colombian government in its Green Growth Strategy (bioeconomy, forestry, sustainable energy) and in the priority of President Duque for the development of creative industries or 'orange economy'. In this context, cooperation on a green recovery from the Cobid-19 crisis deserves particular attention.
- **Investment in rural development and reform** is an important lever for supporting both post-conflict development and decarbonisation in Colombia. Projects financed by the EU Trust Fund have provided valuable support in this area and could be expanded in future. Beyond development cooperation, the EU can also promote European foreign direct investment in the development of climate-resilient agriculture and thereby the sustainable development of Colombia's 'bioeconomy'. Again, the recovery from the Covid-19 crisis may reinforce the related potential.
- **Renewable energy sources beyond hydropower**: Further developing renewable energy sources beyond hydropower can safeguard electricity production when water levels drop in El-Niño years and also help to meet rising energy demand. The EU can support renewable energy development in a variety of ways, for example, by

supporting the development of higher education courses on renewable energy technologies; promoting European investment in Colombian renewable energy projects, through related cooperation on the Covid-19 recovery; and pairing forward-thinking European cities with Colombian cities investing in renewable energy sources.
- **Promotion of just transition narratives in EU (climate) diplomacy with Colombia**: Colombia has to adapt to a decline in its coal exports which the Covid-19 pandemic has accelerated. As well as promoting wider economic diversification, decarbonisation calls for targeted policies to ensure a just transition and new types of employment in Colombia's coal-producing regions. The EU and its member states are developing policies and good practices in this area, as well as gathering lessons learned, and the EU can encourage related bilateral exchange on this issue.
- **Continued cooperation within the UNFCCC**: The EU and Colombia have coordinated their efforts to push for increased ambition within UNFCCC processes. They can continue to work together to push for higher ambition. As implementation advances, countries that have pushed for more ambition will be required to also show they can 'walk the walk', and the EU has significant expertise to support this.

Overall, the EU has already developed a sound basis for cooperation to support both peace and low-carbon development in Colombia, with increasing political dialogue in relevant areas of mutual interest. The EU has proved a valuable partner in the peace process, and it continues to help tackle the root causes of the conflict as a donor, and more and more via investment and trade. The free trade agreement has played a role in shifting EU-Colombia trade relations away from fossil fuels towards agricultural products and, therefore, in supporting rural development and economic diversification in Colombia. This also bodes well for building up trade and other ties (e.g., via higher education and research and development) in other low-carbon, knowledge-based sectors, including in a context of the recovery from the Covid-19 crisis.

Notes

1 An extensive network comprising approximately 3,100 miles of natural gas pipelines transports natural gas throughout the country (EIA 2019). Long self-sufficient, Colombia began regularly importing LNG from the United States for the first time in late 2017 (Kraul 2017). In 2019, LNG imports mainly from Ireland, Nigeria, and the United States valued US$79.7 million (Chatham House 2021).

2 Calculated using World Bank development indicators (GDP in current US$, Merchandise exports in current US$, and fuel exports as percentage of merchandise exports).

3 Venezuela: Assisting refugees affected by Covid-19 in Colombia, https://www.hi-us.org/news_venezuela_assisting_refugees_affected_by_covid_19_in_colombia (visited 21 July 2021).

4 Ecopetrol has been partly privatised, with state-owned companies controlling 89 percent of its shares (Ecopetrol 2018). The largest company in Colombia, Ecopetrol is responsible for producing almost two-thirds of Colombia's oil and owns two of the country's largest refineries and most of the pipeline network (Acosta 2017). The company has begun diversifying into renewable energy (Morais 2018), but the focus is likely to remain heavily on fossil fuels under the new government.

5 'Reducing emissions from deforestation and forest degradation (REDD+) is a mechanism developed by Parties to the United Nations Framework Convention on Climate Change (UNFCCC). It creates a financial value for the carbon stored in forests by offering incentives for developing

countries to reduce emissions from forested lands and invest in low-carbon paths to sustainable development' (UN REDD 2018).

6 The Green Growth Taskforce was initiated by Colombia's National Planning Department. Strategic partners were the Global Green Growth Institute (GGGI), the Government of Norway, the World Bank, the French Development Agency, KfW, UNEP, and the Inter-American Development Bank.

7 A notable exception were the enforceable labour provisions in the agreement, included due to pressure from the European Parliament. The move was intended to ensure that the EU supports the improvement of the labour situation in Colombia, which has one of the world's highest murder rates of trade unionists (Van den Putte and Velluti 2018).

References

Acosta, L.J. (2017) 'Colombia's Ecopetrol bets on non-conventional deposits to boost reserves', *Reuters*, 14 August 2018. Retrieved 31 October 2018 from https://www.reuters.com/article/us-colombia-ecopetrol-idUSKCN1AU1KU.

Acosta, L.J. and Symmes Cobb, J. (2017) 'Private oil companies to invest up to $4.9 billion in Colombia in 2018', *Reuters*, 13 December 2017. Retrieved 29 October 2018 from https://www.reuters.com/article/us-colombia-oil/private-oil-companies-to-invest-up-to-4-9-billion-in-colombia-in-2018-idUSKBN1E72FO?rpc=401&.

Acosta, L.J. and Symmes Cobb, J. (2019) 'Seven companies bid for eleven contracts in Colombia oil auction', *Reuters*, 4 June 2019. Retrieved 26 July 2019 from https://de.reuters.com/article/colombia-oil-auction/seven-companies-bid-for-eleven-contracts-in-colombia-oil-auction-idUKL2N23B109.

Acosta, L.J. and Taj, M. (2019) 'Ecopetrol's plan for fracking project hits new snag in Colombia', *Reuters*, 12 July 2019. Retrieved 26 July 2019 from https://www.reuters.com/article/us-colombia-oil-fracking/ecopetrols-plan-for-fracking-project-hits-new-snag-in-colombia-idUSKCN1U72DM.

Alarcon-Diaz, S. and Lubowski, R. (2018) *Colombia: An Emissions Trading Case Study*. Prepared on behalf of Carbon Trust, EDF and IETA. Available at: https://www.ieta.org/resources/Resources/Case_Studies_Worlds_Carbon_Markets/2018/Colombia-Case-Study-2018.pdf (accessed 29 October 2018).

Amnesty International (2014) '15 Facts about Colombia's Land Restitution Process'. Retrieved 29 October 2018 from https://www.amnesty.org/en/latest/news/2014/11/facts-about-colombia-s-land-restitution-process/.

Baptiste, B., Pinedo-Vasquez, M., Gutierrez-Velez, V.H., Andrade, G.I., Vieira, P., Estupiñán Suárez, L.M., Londoño-Murcia, M., Laurance, W., and Ming Lee, T. (2017) 'Greening peace in Colombia', *Nature Ecology and Evolution*, 1, 0102. https://doi.org/10.1038/s41559-017-0102.

Bellini, E. (2019) 'Colombia launches first RE auction', *Pv Magazine*, 2 January 2019. Retrieved 26 July 2019 from https://www.pv-magazine.com/2019/01/02/colombia-launches-first-re-auction-2/.

Bertelsmann Stiftung (2020) *BTI 2020 country report – Colombia*. Gütersloh: Bertelsmann Stiftung. Available at: https://www.bti-project.org/content/en/downloads/reports/country_report_2020_COL.pdf (accessed 9 July 2021).

Bilotta, N. (2017) 'The FARC, land reform, and the future of Colombia's security', *The Global Risk Insight*, 1 October 2017. Retrieved 29 October 2018 from https://globalriskinsights.com/2017/10/land-farc-future-colombia/.

BP (2020) *BP Statistical Review of World Energy*. Available at: https://www.bp.com/content/dam/bp/business-sites/en/global/corporate/pdfs/energy-economics/statistical-review/bp-stats-review-2021-full-report.pdf (accessed 12 July 2021).

Bristow, M. (2018) '"Peace" brings more murder and cocaine to Colombia's rebel zone', *Bloomberg*, 6 August 2018. Retrieved 29 October 2018 from https://www.bloomberg.com/news/features/2018-08-06/-peace-brings-more-murder-and-cocaine-to-colombia-s-rebel-zones.

Bustos, C. (2017) 'What shapes Colombia's foreign position on climate change?', *Colombia Internacional*, 94, pp. 27–51. https://doi.org/10.7440/colombiaint94.2018.02.

Casey, N. and Abad, S. (2018) 'Colombia elects Iván Duque, a young populist, as president', *The New York Times*, 17 June 2018. Retrieved 31 October 2018 from https://www.nytimes.com/2018/06/17/world/americas/colombia-election-ivan-duque.html.

CAT – Climate Action Tracker (2021) 'CAT Climate Target Update Tracker: Colombia'. Retrieved 23 July 2021 from https://climateactiontracker.org/climate-target-update-tracker/colombia/.

Chaparro, A.C. (2018) 'Exclusive interview: The outlook for EU-Colombia relations in an increasingly divided world', *Vote Watch*, 29 August 2018. Retrieved 31 October 2018 from https://www.votewatch.eu/blog/exclusive-interview-the-outlook-for-eu-colombia-relations-in-an-increasingly-divided-world/.

Chatham House (2021) 'Resource Trade. Earth'. Retrieved 8 July 2021 from http://resourcetrade.earth/.

CIA – Central Intelligence Agency (2021) 'The World Factbook. Colombia'. Retrieved 12 July 2021 from https://www.cia.gov/library/publications/the-world-factbook/geos/co.html.

CIAT (2018) *Boosting Agriculture as Key to Lasting Peace in Colombia*. Retrieved 29 October 2018 from https://ciat.cgiar.org/annual-report-2017-2018/boosting-agriculture-as-key-to-lasting-peace-in-colombia/.

Climatescope (2020) 'Colombia'. Retrieved 8 July 2021 from https://global-climatescope.org/results/CO.

Conservation International (2018) 'Colombia's Carbon Tax Takes Off'. Retrieved 31 October 2018 from https://www.conservation.org/press-releases/2018/04/21/colombias-carbon-tax-takes-off.

DANE (2018) 'Censo Nacional de Población y Vivienda 2018. El Censo Permite al País Ser Más Preciso en la Toma de Decisiones y Orientar Políticas Públicas en Aspectos Sociales'. Retrieved 29 July 2019 from http://www.dane.gov.co/index.php/estadisticas-por-tema/demografia-y-poblacion/censo-nacional-de-poblacion-y-vivenda-2018/cuantos-somos.

Dempsey, C. (2020) 'How foreign investors can benefit from Colombia's Orange Economy?', *The Bogotá Post*, 8 January 2020. Retrieved 9 July 2021 from https://thebogotapost.com/how-foreign-investors-can-benefit-from-colombias-orange-economy/43087/.

DNP – Departamento Nacional de Planeación (2015) 'Compromiso de Reducción de Emisiones de Gases de Efecto Invernadero: Consecuencias Económicas'. Retrieved 10 October 2018 from https://colaboracion.dnp.gov.co/CDT/Estudios%20Econmicos/440.pdf.

DNP (2018) 'Bases del Plan Nacional de Desarrollo 2018–2022 Resumen Ejecutivo'. Retrieved 26 July 2019 from https://www.ambienteysociedad.org.co/wp-content/uploads/2018/11/2018-11-15-Bases-PND-Resumen-Ejecutivo-Completo.pdf.

Duque, I. and Buitrago, F. (2013) *The Orange Economy – An Infinite Opportunity*. IDB – Inter American Development Bank. Available at: https://publications.iadb.org/publications/english/document/The-Orange-Economy-An-Infinite-Opportunity.pdf (accessed 26 July 2019).

Duque, E., González Ruiz, J.D., and Restrepo, J.C. (2017) 'The Clean Development Mechanism as a means to assess the Kyoto Protocol in Colombia', *International Journal of Renewable Energy Research*, 7(3), pp. 1205–1212.

Ecopetrol (2018) 'Ownership'. Retrieved 31 October 2018 from https://www.ecopetrol.com.co/wps/portal/web_es/ecopetrol-web/investors/equity/ownership/.

EEAS – European External Action Service (2018a) *EU-Colombia Relations, Factsheet*. Available at: https://eeas.europa.eu/headquarters/headquarters-homepage_en/16968/EU-Colombia%20relations,%20factsheet (accessed 26 July 2019).

EEAS (2018b) 'Latin America and the Caribbean'. Retrieved 26 July 2019 from https://eeas.europa.eu/regions/latin-america-caribbean/331/latin-america-and-caribbean_en.

EEAS (2019) 'Colombia and the EU Hold a New Edition of Their High Level Political Dialogue'. Retrieved 26 July 2019 from https://eeas.europa.eu/headquarters/headquarters-Homepage/65502/colombia-and-eu-hold-new-edition-their-high-level-political-dialogue_en.

EEAS (2020) 'Joint Press Statement: 6th High-Level Political Dialogue European Union and Colombia'. Retrieved 9 July 2021 from https://eeas.europa.eu/headquarters/headquarters-homepage/90392/joint-press-statement-6th-high-level-political-dialogue-european-union-and-colombia_en.

EIA – US Energy Information Administration (2019) 'Colombia. Analysis'. Retrieved 12 July 2021 from https://www.eia.gov/international/analysis/country/COL/background.

EITI Colombia (2018) 'Colombia'. Retrieved 29 October 2018 from https://eiti.org/es/implementing_country/17.

Energy Policy Tracker (2021) 'Colombia'. Updated 21 July 2021. Retrieved 23 July 2021 from https://www.energypolicytracker.org/country/colombia.

EPRS – European Parliament Research Service (2018) 'Trade agreement between the European Union and Colombia and Peru. European Implementation Assessment'. Retrieved 25 October 2021 from https://www.europarl.europa.eu/RegData/etudes/STUD/2018/621834/EPRS_STU(2018)621834_EN.pdf.

European Commission (2016) 'International Urban Cooperation (IUC)'. Retrieved 26 July 2019 from https://ec.europa.eu/regional_policy/en/policy/cooperation/international/urban/.

European Commission (2015) 'Carta de Intención Sobre un Diálogo en Materia de Política Regional e Integración Fronteriza Entre el Ministerio de Relaciones Exteriores de la República de Colombia y la Dirección General de la Política Regional de la Comisión Europea'. Retrieved 26 July 2019 from https://ec.europa.eu/regional_policy/sources/cooperate/international/pdf/colombia_letter.pdf#page=2.

European Commission (2018) *Commission Staff Working Document – Individual Reports and Info Sheets on Implementation of EU Free Trade Agreements*. Available at: http://trade.ec.europa.eu/doclib/docs/2018/october/tradoc_157473.PDF (accessed 26 July 2019).

European Commission (2021a) 'European Union Trade in Goods with Colombia'. Retrieved 13 July 2021 from https://webgate.ec.europa.eu/isdb_results/factsheets/country/details_colombia_en.pdf.

European Commission (2021b) *Colombia*. Available at: https://webgate.ec.europa.eu/isdb_results/factsheets/country/overview_colombia_en.pdf (accessed 9 July 2021).

FAO – Food and Agriculture Organization of the United Nations (2018) 'Integrating Agriculture in National Adaptation Plans (NAO Ag). Colombia'. Retrieved 31 October 2018 from http://www.fao.org/in-action/naps/partner-countries/colombia/en/.

Felter, C. and Renwick, D. (2017) 'Colombia's civil conflict', *Council on Foreign Relations*, 11 January 2017. Retrieved 29 October 2018 from https://www.cfr.org/backgrounder/colombias-civil-conflict.

Freedom House (2021) 'Colombia'. Retrieved 8 July 2021 from https://freedomhouse.org/country/colombia/freedom-world/2021.

Fund for Peace (2020) 'Colombia's Peace Deal Unravels'. Retrieved 12 July 2021 from https://fragilestatesindex.org/2020/05/10/colombias-peace-deal-unravels/.

Fund for Peace (2021) 'Fragile States Index 2021'. Retrieved 26 July 2021 from https://fragilestatesindex.org/.

Garcia, M. (2016) *EU Trade Relations with Latin America: Results and Challenges in Implementing the EU-Colombia/Peru Trade Agreement*. Prepared on behalf of the European Parliament. Available at: http://www.europarl.europa.eu/RegData/etudes/STUD/2016/534992/EXPO_STU(2016)534992_EN.pdf (accessed 29 October 2018).

GGGI – Global Green Growth Institute (2015) 'Germany, Norway, United Kingdom, and Colombia Announce Agreement to Support Amazon Green Growth'. Retrieved 23 April 2020 from https://gggi.org/press-release/united-kingdom-germany-norway-and-colombia-announce-agreement-to-support-amazon-green-growth/.

GGGI (2018) 'Colombia Approves Long-Term Green Growth Policy and Sets the Course Towards 2030'. Retrieved 31 October 2018 from http://gggi.org/colombia-approves-long-term-green-growth-policy-and-sets-the-course-towards-2030/.

Global Recovery Observatory (2021) 'Colombia'. Retrieved 23 July 2021 from https://recovery.smithschool.ox.ac.uk/tracking/.

Government of Colombia (2015) 'Todos Por Un Nuevo Pais'. Government of Colombia. Retrieved 15 July 2021 from https://observatorioplanificacion.cepal.org/es/planes/plan-nacional-de-desarrollo-todos-por-un-nuevo-pais-de-colombia-2014-2018.

Government of Colombia (2016) 'Final Agreement to End the Armed Conflict and Build a Stable and Lasting Peace'. Retrieved 31 October 2018 from http://especiales.presidencia.gov.co/Documents/20170620-dejacion-armas/acuerdos/acuerdo-final-ingles.pdf.

Government of Colombia (2018) 'Política de Crecimiento Verde'. Retrieved 31 October 2018 from https://www.dnp.gov.co/Crecimiento-Verde/Paginas/Politica-crecimiento-verde.aspx.

Grattan, S. (2018) 'Colombia's rampant corruption a hot topic in presidential vote', *Reuters*, 15 June 2018. Retrieved 29 October 2018 from https://www.reuters.com/article/us-colombia-election-corruption/colombias-rampant-corruption-a-hot-topic-in-presidential-vote-idUSKBN1JB252.

Green Climate Fund (2018) 'Colombia'. Retrieved 22 April 2020 from https://www.greenclimate.fund/countries/colombia.

GRICCE – Grantham Research Institute on Climate Change and the Environment (2016) 'National Climate Change Decree (SISCLIMA)'. Retrieved 31 October 2018 from http://www.lse.ac.uk/GranthamInstitute/law/national-climate-change-decree-sisclima/.

GRICCE (2021) 'Colombia'. Retrieved 23 July 2021 from https://climate-laws.org/geographies/colombia.

Grieger, G. and Harte, R. (2018) *EU Trade with Latin America and the Caribbean*. Prepared on behalf of the European Parliament. Available at: http://www.europarl.europa.eu/RegData/etudes/IDAN/2018/625186/EPRS_IDA(2018)625186_EN.pdf (accessed 29 October 2018).

ICAP (2021) 'ETS Map. Colombia'. Retrieved 9 July 2021 from https://icapcarbonaction.com/en/ets-map?etsid=92.

IEA – International Energy Agency (2017a) *Coal 2017. Analysis and Forecasts to 2022. Market Report Series*. Organisation for Economic Co-operation and Development, and International Energy Agency, Paris. Available at: https://iea.blob.core.windows.net/assets/9f9f4b45-66c4-4790-a16f-7cf5799b8fdf/Coal2017.pdf (accessed 15 July 2021).

IEA (2017b) 'Energy Plan 2050 (Plan Energético Nacional Colombia: Ideario Energético 2050)'. Retrieved 26 July 2019 from https://www.iea.org/policies/6297-energy-plan-2050-plan-energetico-nacional-colombia-ideario-energetico-2050.

IEA (2020) 'Colombia. Electricity Generation by Source'. Retrieved 8 July 2021 from https://www.iea.org/countries/colombia.

IETA (2018) *Colombia: An Emissions Trading Case Study*. Available at: https://www.ieta.org/resources/Resources/Case_Studies_Worlds_Carbon_Markets/2018/Colombia-Case-Study-2018.pdf (accessed 26 July 2019).

IHA – International Hydropower Association (2018) 'Colombia'. Retrieved 29 October 2018 from https://www.hydropower.org/country-profiles/colombia.

IKI – Internationale Klimaschutzinitiative (2018) 'Strategien ökosystembasierter Anpassung an den Klimawandel in Kolumbien und Ecuador'. Retrieved 23 April 2020 from https://www.international-climate-initiative.com/de/details/project/strategien-oekosystembasierter-anpassung-an-den-klimawandel-in-kolumbien-und-ecuador-12_II+_003-476/.

IMF – International Monetary Fund (2018) 'A Tax Reform Succeeds in Colombia'. Retrieved 29 October 2018 from https://www.imf.org/external/mmedia/view.aspx?vid=5766970768001.

International Crisis Group (2019a) 'Watch List 2019'. Retrieved 26 July 2019 from https://reliefweb.int/sites/reliefweb.int/files/resources/watch-list-update-july-2019.pdf.

International Crisis Group (2019b) 'Crucial Reforms Languish as Colombia Seeks to Consolidate Peace'. Retrieved 26 July 2019 from https://www.crisisgroup.org/latin-america-caribbean/andes/colombia/crucial-reforms-languish-colombia-seeks-consolidate-peace.

International Crisis Group (2020) 'Colombia: Peace Withers amid the Pandemic'. Retrieved 23 July 2021 from https://www.crisisgroup.org/latin-america-caribbean/andes/colombia/colombia-peace-withers-amid-pandemic.

Ioannides, I. (2019) *Peace and Security in 2019: Evaluating EU Efforts to Support Peace in Colombia*. Brussels: European Parliamentary Research Service. Available at: https://www.europarl.europa.eu/RegData/etudes/STUD/2019/631746/EPRS_STU(2019)631746_EN.pdf (accessed 23 July 2021).

IRENA – International Renewable Energy Agency (2021a) 'Renewable Capacity Statistics 2021'. Retrieved 8 July 2021 from https://www.irena.org/publications/2021/March/Renewable-Capacity-Statistics-2021.

IRENA (2021b) 'Renewable Energy Auctions in Colombia: Context, Design and Results'. Retrieved 9 July 2021 from https://www.irena.org/publications/2021/March/Renewable-energy-auctions-in-Colombia.

Kraul, C. (2017) 'Colombia begins importing bulk LPG as natural gas reserves, output fall', *S&P Global*, 15 December 2017. Retrieved 29 October 2018 from https://www.spglobal.com/platts/en/market-insights/latest-news/natural-gas/121517-colombia-begins-importing-bulk-lpg-as-natural-gas-reserves-output-fall.

Kuramochi, T., Nascimento, L., de Villafranca Casas, M.J., Fekete, H., de Vivero, G., Lui, S., Kurdziel, M., Moisio, M., Tanguy, P., Jeffery, L., Schiefer, T., Suzuki, M., Höhne, N., van Soest, H., den Elzen, M., Esmeijer, K., Roelfsema, M., Forsell, N., and Gusti, M. (2019) *Greenhouse Gas Mitigation Scenarios for Major Emitting Countries. Analysis of Current Climate Policies and Mitigation Commitments: 2019 Update*. New Climate Institute. Available at: https://www.pbl.nl/sites/default/files/downloads/pbl-2019-greenhouse-gas-mitigation-scenarios-for-major-emitting-countries-2019-update_3626.pdf (accessed 9 July 2021).

LSE – London School of Economics and Political Science (2019) 'Ministerio de Ambiente y Desarrollo Sostenible – Decreto No. 298 2016'. Retrieved 26 July 2019 from http://www.lse.ac.uk/GranthamInstitute/wp-content/uploads/laws/2046.pdf.

Martínez, C. and Alfonso, W.H. (2018) 'Climate change in Colombia: A study to evaluate trends and perspectives for achieving sustainable development from society', *International Journal of Climate Change Strategies and Management*, 10(4), pp. 632–652.

Ministry of Mines (2018) 'Energia Para Todos'. Retrieved 26 July 2019 from https://twitter.com/MinMinas/status/1039281463613636609.

Monge, C. (2018) 'Colombia puts a tax on carbon', *Conservation Finance Network*, 27 November 2018. Retrieved 26 July 2019 from https://www.conservationfinancenetwork.org/2018/11/27/colombia-puts-tax-on-carbon.

Morais, L. (2018) 'Colombian oil company Ecopetrol to build 1st PV plant in 2019', *Renewables Now*, 20 August 2018. Retrieved 31 October 2018 from https://renewablesnow.com/news/colombian-oil-company-ecopetrol-to-build-1st-pv-plant-in-2019-624070/.

Murphy, H. and Acosta, L.J. (2018) 'Colombia's President-elect names Maria Fernanda Suarez energy minister', *Reuters*, 18 July 2018. Retrieved 29 October 2018 from https://www.reuters.com/article/us-colombia-cabinet-miningenergy/colombias-president-elect-names-maria-fernanda-suarez-energy-minister-idUSKBN1K81RW.

Natural Resources Governance Institute (2021) 'Colombia: Updated Assessment of the Impact of the Coronavirus Pandemic on the Extractive Sector and Resource Governance'. Retrieved 23 July 2021 from https://resourcegovernance.org/analysis-tools/publications/colombia-updated-assessment-impact-coronavirus-extractive.

ND-GAIN – Notre Dame Global Adaptation Initiative (2020) 'ND-GAIN Country Index'. South Bend, IN: University of Notre Dame. Retrieved 5 July 2021 from https://gain.nd.edu/our-work/country-index/.

NDC Cluster (2018) 'Colombia'. Retrieved 31 October 2018 from https://www.ndc-cluster.net/country/colombia.

NDC Partnership (2017) *NDC Country Outlook. Colombia*. Available at: https://ndcpartnership.org/sites/all/themes/ndcp_v2/docs/country-engagement/countries/NCDP_Outlook_Colombia_v6a.pdf (accessed 29 October 2018).

Oxford Business Group (2018) 'Peace Deal and Agro-Industry Development Allow Colombia to Realise its Agricultural Potential'. Retrieved 29 October 2018 from https://oxfordbusinessgroup.com/overview/ready-grow-new-peace-deal-will-allow-sector-realise-its-potential.

Oxford Business Group (2019) 'Agri-Business Opportunities in Colombia Continue to Grow'. Retrieved 8 July 2021 from https://oxfordbusinessgroup.com/overview/space-grow-number-developments-and-efforts-raise-financial-capacity-bode-well-sector-expansion-and.

OECD – Organization for Economic Cooperation and Development (2014) 'Colombia Environmental Performance Review'. Retrieved 29 October 2018 from http://www.oecd.org/environment/country-reviews/oecd-environmental-performance-reviews-colombia-2014-9789264208292-en.htm.

OECD (2016) 'Education in Colombia'. Retrieved 29 October 2018 from http://www.oecd.org/education/school/Education-in-Colombia-Highlights.pdf.

OECD (2017) 'Colombia: Raise Productivity and Make Growth More Inclusive'. Retrieved 29 October 2018 from http://www.oecd.org/newsroom/colombia-raise-productivity-and-make-growth-more-inclusive.htm.

OECD (2018a) 'A Mutually Beneficial Relationship'. Retrieved 29 October 2018 from http://www.oecd.org/latin-america/countries/colombia/.

OECD (2018b) 'The New OECD Jobs Strategy. Good Jobs for All in a Changing World of Work. How Does Colombia Compare?'. Retrieved 8 July 2021 from https://www.oecd.org/colombia/jobs-strategy-COLOMBIA-EN.pdf.

OECD (2021) 'Colombia Economic Snapshot. Economic Forecast Summary (May 2021)'. Retrieved 8 July 2021 from https://www.oecd.org/economy/colombia-economic-snapshot/.

Oei, P.Y. and Mendelevitch, R. (2016) *Perspectives on Colombian Coal Exports on the International Steam Coal Market Until 2030*. Prepared on behalf of the Rosa Luxemburg Stiftung. Available at: https://www.rosalux.de/fileadmin/rls_uploads/pdfs/engl/Onlinestudie_Kohle_en.pdf (accessed 29 October 2018).

OXFAM (2016) 'Unearthed: Land, Power and Inequality in Latin America'. Retrieved 29 October 2018 from https://www.oxfam.org/en/research/unearthed-land-power-and-inequality-latin-america.

Paez, A.F., Muñoz Maldonado, Y., Ospino Castro, A., Hernandez, N., Cande, E., Paecheco, L., Gonzales, W. and Sotelo, O. (2017) 'Future scenarios and trends of energy demand in Colombia using long-range energy alternative planning', *International Journal of Energy Economics and Policy*, 7(5), pp. 178–190.

Palmer, L. (2017) '"It's a perverse system": How Colombia's farmers are reforesting their logged land', *The Guardian*, 29 December 2017. Retrieved 31 October 2018 from https://www.theguardian.com/environment/2017/dec/29/its-a-perverse-system-how-colombias-farmers-are-reforesting-their-logged-land.

Peñaloza, M. (2019) 'Colombia: The National Development Plan and the environment', *Latinamerican Post*, 3 April 2019. Retrieved 26 July 2019 from https://latinamericanpost.com/27328-colombia-the-national-development-plan-and-the-environment.

ProColombia (2018) 'An Overview of Foreign Direct Investment in Colombia'. Retrieved 31 October 2018 from http://www.investincolombia.com.co/news/921-an-overview-of-foreign-direct-investment-in-colombia.html.

Reardon, S. (2018) 'Peace is killing Colombia's jungle – and opening it up', *Nature*, 558, pp. 169–170. https://doi.org/10.1038/d41586-018-05397-2.

Reith, S. (2021) 'Five Years of the Colombian Peace Agreement'. Available at: https://www.kas.de/en/web/auslandsinformationen/artikel/detail/-/content/zwischenbilanz-mit-licht-und-schatten (accessed 23 July 2021).

Semana (2015) 'Alta Tensión en el Sector Eléctrico'. Retrieved 29 September 2018 from https://www.semana.com/economia/articulo/sector-electrico-en-colombia-afectado-por-el-fenomeno-de-el-nino/447302-3.

Slav, I. (2017) 'Colombia looks at shale as oil reserves plunge', *Oil Price.com*, 21 August 2017. Retrieved 29 October 2018 from https://oilprice.com/Energy/Crude-Oil/Colombia-Looks-At-Shale-As-Oil-Reserves-Plunge.html.

Smith, M. (2018) 'Colombia's oil industry faces 3 big problems', *Oil Price.com*, 13 February 2018. Retrieved 29 October 2018 from https://oilprice.com/Energy/Crude-Oil/Colombias-Oil-Industry-Faces-3-Big-Problems.html.

Strambo, C. (2017) 'Transition Lessons from Colombian Coal Politics'. SEI: Stockholm Environment Institute. Retrieved 31 October 2018 from https://www.sei.org/perspectives/colombian-coal-politics/.

Strambo, C. and Atteridge, A. (2018) *How Colombia Can Plan for a Future without Coal*. Prepared on behalf of SEI: Stockholm Environment Institute. Available at: https://www.sei.org/wp-content/uploads/2018/08/how-colombia-can-plan-for-a-future-without-coal.pdf (accessed 29 October 2018).

Statista (2021) 'Colombia: Distribution of Employment by Economic Sector from 2010 to 2020'. Retrieved 9 July 2021 from https://www.statista.com/statistics/369122/employment-by-economic-sector-in-colombia/.

Trademap (2021) 'Trade Statistics for International Business Development'. Retrieved 8 July 2021 from https://www.trademap.org/Index.aspx.

Trading Economics (2021) 'Colombia Unemployment Rate'. Retrieved 23 July 2021 from https://tradingeconomics.com/colombia/unemployment-rate.

Transparency International (2021) 'Corruption Perceptions Index 2020'. Retrieved 8 July 2021 from https://www.transparency.org/en/cpi/2019/index/col.

UNDP – United Nations Development Programme (2020) 'Human Development Report'. Available at: http://hdr.undp.org/en/countries (accessed 26 July 2021).

UNEP – United Nations Environment Programme (2016) 'The Environmental Dimension of Peace. Delivering on the 2030 Agenda in Colombia'. Retrieved 29 October 2018 from http://web.unep.org/ourplanet/may-2016/articles/environmental-dimension-peace.

UNEP (2017) 'Environmental Recovery Key to Post-Conflict Development in Colombia – UN Agency'. Retrieved 29 October 2018 from https://news.un.org/en/story/2017/03/553942-environmental-recovery-key-post-conflict-development-colombia-un-agency.

UNHCR – United Nations High Commissioner for Refugees (2020) 'Global Focus. UNHCR Operations Worldwide. Colombia. 2020 Year-End Report'. Retrieved 8 July 2021 from https://reporting.unhcr.org/colombia.

UNHCR (2021) 'Colombia Operational Update'. May 2021. Retrieved 8 July 2021 from https://reporting.unhcr.org/sites/default/files/Colombia%20Operational%20Update%20May%202021.pdf.

UN REDD (2018) 'About REDD+'. Retrieved 29 October 2018 from https://www.unredd.net/about/what-is-redd-plus.html.

US AID (2017) *Fact Sheet. Climate Risk Profile Colombia*. Available at: https://www.climatelinks.org/sites/default/files/asset/document/2017_USAID%20CCIS_Climate%20Risk%20Profile_Colombia.pdf (accessed 29 October 2018).

Van den Putte, L. and Velluti, S. (2018) 'The promotion of social trade by the European Union in its external trade relations'. In: Sangeeta, K. and García, M. (eds.) *Handbook on the EU and International Trade*. Cheltenham: Edward Elgar Publishing, pp. 224–242.

World Bank (2020a) 'Human Capital Project. Colombia. Human Capital Index 2020'. Retrieved 9 July 2021 from https://databank.worldbank.org/data/download/hci/HCI_2pager_COL.pdf?cid=GGH_e_hcpexternal_en_ext.

World Bank (2020b) 'Ease of Doing Business in Colombia'. Retrieved 9 July 2021 from https://www.doingbusiness.org/en/data/exploreeconomies/colombia.

World Bank (2021a) 'Poverty Headcount Ratio at $1.90 a Day (2011 PPP) (% of Population) – Colombia'. Retrieved 8 July 2021 from https://data.worldbank.org/indicator/SI.POV.DDAY?locations=CO.

World Bank (2021b) 'Poverty Headcount Ratio at National Poverty Lines (% of Population) – Colombia'. Retrieved 8 July 2021 from https://data.worldbank.org/indicator/SI.POV.NAHC?locations=CO.

World Bank (2021c) 'Fuel Exports (% of Merchandise Exports) – Colombia'. Retrieved 8 July 2021 from https://data.worldbank.org/indicator/TX.VAL.FUEL.ZS.UN?locations=CO.

World Bank (2021d) 'GDP (Current US$) – Colombia'. Retrieved 9 July 2021 from https://data.worldbank.org/indicator/NY.GDP.MKTP.CD?locations=CO.

World Bank (2021e) 'Worldwide Governance Indicators 2019'. Retrieved 26 July 2021 from http://info.worldbank.org/governance/wgi/#reports.

World Bank (2021f) 'Government Expenditure on Education, Total (% of GDP)'. Retrieved 9 July 2021 from https://data.worldbank.org/indicator/SE.XPD.TOTL.GD.ZS?locations=CO-OE-EU.

World Bank (2021g) *Regulatory Indicators for Sustainable Energy (RISE)*. Retrieved 26 July 2021 from https://rise.esmap.org.

World Population Review (2021) 'Colombia Population 2021'. Retrieved 8 July 2021 from http://worldpopulationreview.com/countries/colombia-population/.

WRI – World Resources Institute (2021) 'Climate Watch. Colombia. Greenhouse Gas Emissions and Emission Targets'. Retrieved 12 July 2021 from https://www.climatewatchdata.org/countries/COL.

WWF (2017) 'Colombian Congress Passes Bill Ratifying Paris Agreement'. Retrieved 31 October 2018 from http://www.wwf.org.co/?303172/Colombia-ratifying-Paris-Agreement.

Yagoub, M. (2016) 'How decentralizing political power strengthened Colombia crime groups', *InSight Crime*, 2 November 2016. Retrieved 31 October 2018 from https://www.insightcrime.org/news/analysis/how-decentralizing-political-power-strengthened-colombia-crime-groups/.

6 Qatar

Moving beyond oil and gas within a fragile regional context

Introduction

The State of Qatar (henceforth Qatar) is a country of around 2.8 million inhabitants (UN DESA 2019) located in the Persian Gulf with neighbouring countries Saudi Arabia, the United Arab Emirates, Kuwait, and Iran. Of the total population, approximately 2 million are foreign workers, mainly from India, Nepal, Bangladesh, and the Philippines. When Qatar started exploiting its fossil fuel resources in 1949, it had less than 16,000 inhabitants. In the twenty-first century, population growth has been driven primarily by an increase of (primarily male) foreign workers (De Bel-Air 2014). Qatar has a territory of only 11,571 km^2 (somewhat less than Northern Ireland) and is composed of eight municipalities, with Doha as the country's biggest city and capital. The country is highly urbanised with 99.3 percent of the inhabitants living in cities in 2021 (CIA 2021). Qatar gained independence from the United Kingdom in 1971 and is ruled as an absolute monarchy by Emir Tamim bin Hamad Al Thani, who has been in power since 2013.

In recent decades, Qatar has experienced dynamic economic development, driven by its oil and gas industry. From 1995 to 2019, Qatar's gross domestic product (GDP) increased more than 20-fold to about US$175.8 billion (in current US$) (World Bank 2021a). Over the same period, GDP per capita grew nearly four-fold to about US$62,000 (World Bank 2021b), one of the highest per capita incomes in the world. By 2017, industry and services each contributed approximately half of Qatar's GDP, while the role of agriculture was negligible (0.2 percent) (CIA 2021). Poverty is an issue to some extent among the foreign population, but unemployment is all but absent. Qatar has been selected as the host of the 2022 FIFA World Cup, a prospect that has motivated numerous ambitious construction projects in the country (which have drawn significant criticism – see below). Overall, Qatar has become a high-income country, ranking relatively high on economic welfare and human development indices.

Qatar's relations with other countries in the region have been under strain in the 2010s, and particularly after 2017 when several neighbouring countries led by Saudi Arabia cut off diplomatic relations and established an economic embargo against Qatar. They accused Qatar of supporting the Muslim Brotherhood and of being a close ally of Iran. The embargo constituted an enormous challenge, since Qatar was highly dependent on its neighbouring countries for food imports and general trade. In 2016, Qatar had imported goods and services valued US$5 billion from the countries behind the embargo

DOI: 10.4324/9781003183037-6

and exported about US$5.6 billion of goods and services to these countries. However, under the embargo Qatar largely succeeded in rebuilding its trade links and food supply chains with other countries (Collins 2018; CIA 2021). On 1 January 2019, Qatar furthermore withdrew from the Organization of the Petroleum Exporting Countries (OPEC), a sign of the troubled relationship with Saudi Arabia and its allies as well as of its strategic economic orientation towards producing and exporting natural gas rather than oil. The embargo by Saudi Arabia and others was eventually lifted in January 2021.

In line with the overall approach of the case studies, this chapter will explore the challenges and opportunities Qatar and Qatar-EU relations are facing under decarbonisation in five steps. The next section first substantiates that Qatar is highly dependent on the production and export of oil and gas and hence vulnerable to price fluctuations and the phase-out of fossil fuels. It also analyses other fragility and security risks, including the geopolitical position of Qatar in the Gulf region and the Middle East more generally. Subsequently, the status of Qatar's climate policy is assessed, which leaves significant room for further improvement. This is followed by a review of relevant trends and potentials, especially regarding efforts to diversify the economy that have only been effective to a limited extent so far. The one but last section reviews the framework and basis of EU-Qatar relations beyond fossil fuels. The final section concludes and identifies focal points for EU foreign policy towards Qatar under decarbonisation.

Exposure and risk

Exposure to global decarbonisation trends

Qatar's economy is highly dependent on the production and export of oil and gas. It is the largest exporter of liquefied natural gas (LNG) in the world, which accounts for a large share of its GDP and government revenue. Qatar consequently also remains exposed to oil price fluctuations, and investments in oil and gas may become 'stranded'.

Qatar is a significant producer of gas and oil. In 2019, it produced about 1.88 million barrels of oil per day, equivalent to about 2 percent of the world total. Qatar's gas production reached 178 billion cubic metres in 2019 (nearly a doubling over the previous decade), which represented 4.5 percent of global production (BP 2020). There is no coal production in the country (IEA 2015).

Qatar possesses both significant oil and gas reserves, with the latter being particularly impressive. Its proven oil reserves are estimated to amount to around 25 billion barrels, which could support current production levels for about another 35 years (BP 2020). Estimated gas reserves are even more abundant and amount to 25 trillion cubic metres, the third largest gas reserves in the world after Russia and Iran (BP 2020; see for other estimates: EIA 2015; CIA 2021). They could support current production levels for nearly 140 years. Gas production and exports are expected to increase further (Meltzer et al. 2014).

In accordance with its resource base, Qatar mainly uses gas and oil to supply its own energy needs. Since 2009, its oil consumption has doubled to about 346,000 barrels per day in 2019 (close to 20 percent of production) (BP 2020). Similarly, gas consumption has been steadily and steeply increasing, nearly doubling to over 41 billion cubic metres in 2019 (equivalent to about 23 percent of production) (BP 2020). Overall, about three quarters of overall energy consumption is accounted for by gas and the remaining quarter by oil (BP 2020).[1] Plans to diversify the energy mix and develop renewable energy sources have yet to bear fruit (see also below).

Exports of oil and gas are of crucial importance for Qatar's economy (see Figure 6.1). At the end of the 2010s, the country exported more than 80 percent of its oil production and more than 75 percent of its gas production (see above). Oil and gas account for nearly 90 percent of Qatari merchandise exports, with this share having fluctuated somewhat over the past two decades (varying between 70 and 95 percent), also as a result of price fluctuations (World Bank 2021i). Qatar is the world's largest exporter of LNG (BP 2020). Consequently, oil and gas play a dominant role in Qatar's economy and government budget. The oil and gas industries are the engines of the economy, with the sectors accounting for around 21 percent of GDP in 2019 (fluctuating somewhat with the oil price) and for the lion's share of the state budget (World Bank 2021e, 2021f). To be sure, Qatar has made significant efforts to diversify its economy. Notably, the Emir Hamad bin Khalifa

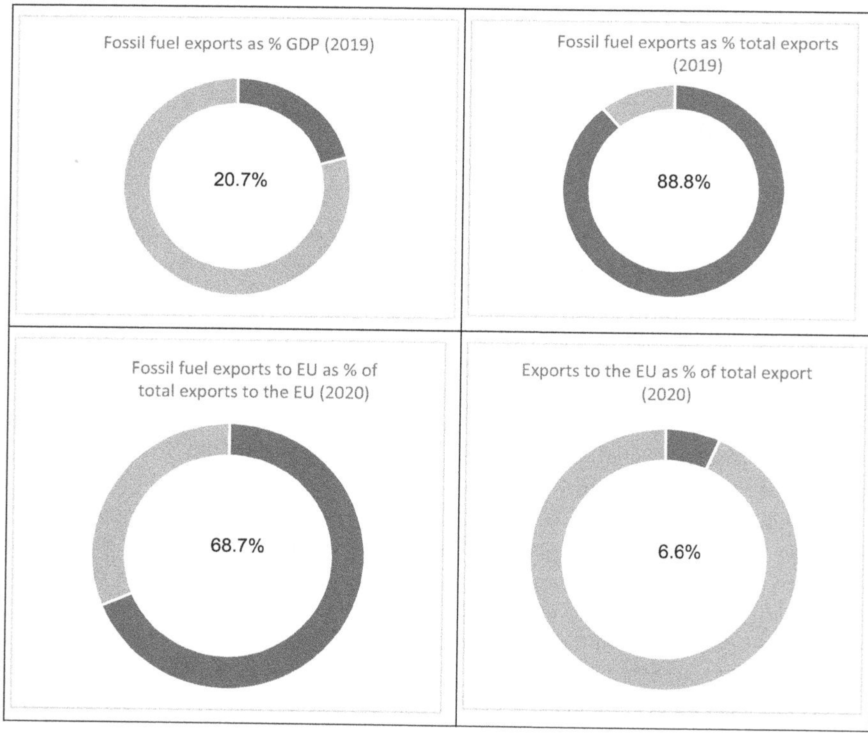

Figure 6.1 Status of fossil fuel exports of Qatar in 2019/2020.

Sources: Data based on https://data.worldbank.org (fossil fuel exports as a percentage of merchandise export; oil, gas, and coal as a percentage of GDP – accessed on 25 July 2021) and European Commission (2021) (exports to EU – accessed on 25 July 2021).

al Thani founded the Qatar Investment Authority in 2005 to invest revenues from the oil and gas sectors in a variety of projects in Qatar and abroad. The Investment Authority makes investments in international markets (including in Europe) and in non-oil projects in Qatar itself (Qatar Investment Authority 2016). Even though the share of non-oil/gas sources of government revenues grew to 22 percent in 2018 (Federal Reserve Bank of St. Louis 2018), the fossil fuel sector still generates the largest share of government revenues and Qatar remains highly dependent on these sources of income. Indeed, a previous diversification strategy was to increase gas production and exports (in order to diversify away from oil).

Qatar's government also provides significant subsidies for energy. According to the International Energy Agency (IEA), the country supported oil, gas, and electricity with about US$500 million in 2020. However, this constituted a major decrease from subsidies of more than US$4 billion in 2014 (IEA 2020), when Qatar was among the countries with the highest energy subsidies per capita in the world (with electricity being free for Qatari nationals: Meltzer et al. 2014).

Qatar remains highly exposed to variations in oil and gas prices. Declining oil prices after 2014 and reduced demand in the wake of the Covid-19 crisis have left clear marks on economic development. GDP dropped from US$206 billion in 2014 to 151.7 billion in 2016 and from US$183.3 billion in 2018 to US$146.4 billion in 2020 (in current US$). GDP per capita similarly dropped from nearly US$84,000 in 2014 to about US$57,000 in 2016 and from nearly US$66,000 in 2018 to about US$51,000 in 2020 (World Bank 2021a, 2021b). Due to exchange rate fluctuations, the figures differ in constant 2010 US$, according to which overall GDP remained at above US$170 billion from 2015 to 2020, whereas GDP per capita decreased from about US$67,000 in 2014 to about US$60,000 in 2020 (World Bank 2021c, 2021d). Given the robust demand for gas, the conditions for a relatively strong recovery of Qatar's economy from the Covid-19 crisis seem to be in place (World Bank 2021h).

Accordingly, the government budget situation has also seen significant fluctuations. Qatar's government revenues increased from 54 billion Qatari Rials in 2005 to 360.6 billion Qatari Rials in 2014, before decreasing, in the wake of plummeting oil prices, to 185.4 billion Qatari Rials in 2018 (equivalent to about €42 billion). As oil prices recover and gas production increases, government revenues are expected to increase (Quandl 2021). Qatar's public debt-to-GDP ratio rose from 32.3 percent in 2014 to 71.8 percent in 2020. The increase was only temporarily interrupted by the recovery of oil prices, with the ratio dropping to 48.4 percent in 2018 (Trading Economics 2021). Overall, the debt-to-GDP ratio has remained modest compared to the international average, even though it remains vulnerable to oil price fluctuations (with the full impact of the Covid-19 pandemic still materialising).

Reflecting its resource base, Qatar has a vast network of pipelines, transport ships, and extraction and production facilities for oil and gas. Petroleum and petroleum products are exported from three major terminals: Umm Said (Mesaieed), Halul Islan, and Ras Laffan. The oil pipeline network of the state-owned company Qatar Petroleum (see below) brings oil from offshore fields to the Halul Islan terminal where it is further processed and then exported. Oil from onshore sources is mainly transported to the Umm Said (Mesaieed) terminal where the oil is either refined or exported. Qatar is also a world leader in gas-to-liquids technologies and has two operational facilities that can turn dry natural gas into liquid fuels (EIA 2015).

The oil and gas sector has remained a major focus of Qatari investments, creating a high risk of stranded assets under decarbonisation. The emphasis on fossil fuel investments has increasingly shifted towards gas (which may somewhat mitigate the decarbonisation risk at least in the medium term). In 2018, Qatar announced investments of US$12 billion in the oil and gas sectors, out of a total of new investments of US$85 billion in different sectors. The three biggest planned projects are the Bul Hanine Field Redevelopment Project (US$11 billion), the North Field Gas Development Project (US$2 billion), and the Barzan Gas Development Offshore Project (US$700 million) (Trade Arabia 2018). The North Field is estimated to hold recoverable gas reserves of more than 900 trillion standard cubic feet, which makes it the largest single non-associated gas reservoir in the world (Offshore Energy Today 2017). The Barzan Gas Development Offshore Project forms part of the broader Barzan Gas Project, a joint venture between Qatar Gas and Exxon Mobile. This US$10.4 billion project includes onshore and offshore developments, a gas processing unit, and several pipelines (EIA 2015; Hydrocarbons Technology 2018). Several of the investment projects involve foreign investors (such as Exxon Mobile, Royal Dutch/Shell, and Total; Zhdannikov 2018).

Other fragility and security risks

Qatar also faces a number of other risks. To start with, whereas Qatar has developed into an advanced economy and has low levels of corruption or repression, it does face challenges with respect to democracy and political rights. Ranking 144 out of 178 countries on the Fragile States Index in 2021, Qatar can be considered relatively stable, although sub-indicators do indicate issues with respect to fluctuating economic inequality and state legitimacy (Fund for Peace 2021; see also Figures 6.2 and 6.3). Corruption is not a major concern, with the country ranked 30 among the 180 countries on the Transparency International Corruption Perceptions Index (Transparency International 2021). The Freedom House Index of 2021 categorises Qatar as 'not free' with problems indicated in particular for general freedom, political rights, and civil liberties (Freedom House 2021). This is in line with the findings of the Bertelsmann Stiftung's Transformation Index, which

Fragility (2021)	Sustainable	**STABLE**	Warning	Alert
Human development (2019)	**VERY HIGH**	High	Medium	Low
Strength of governance (2019)	Very high	**HIGH**	Medium	Low
Climate change vulnerability (2019)	Low	**MEDIUM**	High	Very high
Sustainable energy development (2019)	Very high	**HIGH**	Medium	Low

Figure 6.2 Governance indicators (Qatar).

Sources: Fund for Peace (2021), ND-GAIN (2020), UNDP (2020), World Bank (2021j, 2021k).

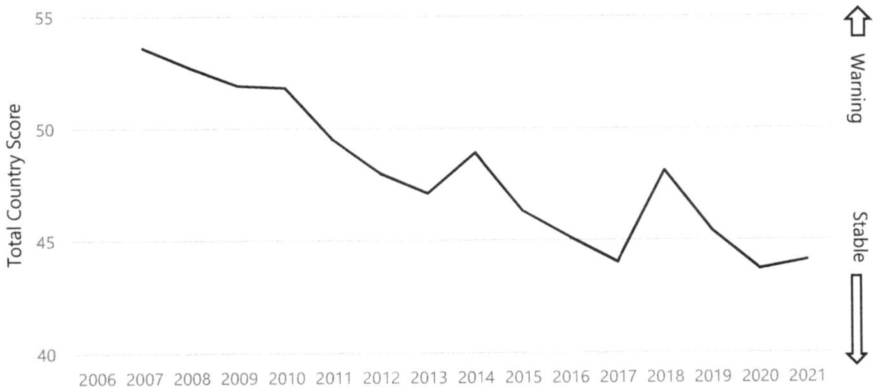

Figure 6.3 Qatar's performance in the Fragile States Index 2006–2021.
Source: Fund for Peace (2021).
Note: Higher index numbers indicate a higher level of state fragility.

shows a clear contrast between strong progress towards establishing a market economy and general economic performance, on the one side, and clear deficiencies with respect to political participation, rule of law, and democratic institutions, on the other (Bertelsmann Stiftung 2020).

A particular challenge for Qatar is the highly problematic situation of the about 2 million expatriate workers, mainly low-skilled Asian construction workers. While Qatar's sizeable public sector employs about 80 percent of Qatari nationals, almost 80 percent of non-nationals are employed in the private sector, mainly in the construction sector. While the numbers of expatriate workers have increased, Qatar has been criticised for the exploitation of these blue-collar workers especially in the construction sector, including in preparation for the 2022 FIFA World Cup (Amnesty International 2020). The very large number of foreign workers has led to concerns about their relationship with native Qataris and the long-term prospects for their future and integration into Qatari society. According to its National Vision for 2030, Qatar is aiming to attract more high-skilled workers (De Bel-Air 2014).

In addition, the Middle East and the Gulf region provide for a sensitive and challenging security environment. The region has long been subject to serious political tensions, and these have repeatedly culminated in violent and armed conflict (most recently in Syria and Yemen). The rivalry between Iran and Saudi Arabia for regional dominance is a key factor shaping the region's precarious security situation, as is the engagement of the United States and Russia (as well as other players such as Turkey). Other countries and actors have strategically aligned themselves with these major players according to the resulting fault lines in various ways.

The escalation of tensions between Saudi Arabia and others, on the one hand, and Qatar on the other, in the 2010s, is one expression and result of this precarious regional security balance. Saudi Arabia, the United Arab Emirates, and Bahrain first withdrew their ambassadors from Doha in 2014. One of the reasons given was that Qatar did not want to categorise the Muslim Brotherhood as a terrorist organisation. In 2017, the crisis escalated further. Qatar was asked, among other things, to cut its ties with the Muslim Brotherhood, loosen its relations with Iran, and shut down the Al Jazeera news network. When Qatar refused to follow these requests, Saudi Arabia and its allies launched an embargo against Qatar (as discussed in the introduction), throwing the Gulf Cooperation Council (GCC) into paralysis (Galeeva 2018). In January 2021, the embargo was lifted on the basis of an agreement reached during a GCC summit with the mediation of Kuwait and the United States (Salam and Alam 2021).

With respect to climate change impacts, these are likely to be significant in Qatar. In particular, Qatar is highly likely to be affected by sea-level rise and the associated risk of salinisation of groundwater. With water demand increasing due to industrial uses and population growth, water scarcity is already an issue, and precipitation is forecast to decrease with climate change. Rising sea levels might require moving much of Qatar's population (of which 96 percent live in coastal towns or cities; Qatar 2011) away from coastal areas and could endanger oil and gas as well as power-producing infrastructure. Further increases of already very high temperatures also pose challenges (Qatar 2011; Meltzer et al. 2014; Al-Sarihi 2018).

Nevertheless, Qatar ranked low on the Global Climate Risk Index for 2019 and for the period 2000–2019. The Index assesses the extent to which countries have been subject to weather-related loss events (storms, floods, heat waves, etc.) (Germanwatch 2021). Similarly, the index of the Notre Dame Global Adaptation Initiative (ND-GAIN), which considers eight life-supporting sectors (food, water, health, ecosystems, human habitat, coastal, energy, and transportation infrastructure), gives Qatar a favourable medium score and rank for its overall vulnerability and readiness. Its vulnerability is considered less than average and its readiness slightly higher than average (ND-GAIN 2020; see Figure 6.2). Hence, climate change impacts in Qatar are considered significant but modest in international comparison.

Past and present efforts to decarbonise

Qatar can be considered a climate policy laggard. It is a party to all major climate treaties. It ratified the UN Framework on Climate Change (UNFCCC) in 1996, the Kyoto Protocol in 2005, and the Paris Agreement in 2017. It is a non-Annex I party to the UNFCCC and hence – despite its status as a high-income country – considered a developing country for the purposes of the climate treaties. Within the UNFCCC negotiations, Qatar forms part of the Arab States, a group (led by Saudi Arabia) that is widely considered as 'obstructionist' in the international climate negotiations. Arab oil producers see climate policy and its international cooperation as a serious threat to their economic welfare and political stability (Coates Ulrichsen 2010; Reiche 2010; Carbon Brief 2015a). Hence, Qatar does not play a progressive role in international climate policy.

Consequently, Qatar's first Nationally Determined Contribution (NDC) submitted under the Paris Agreement in 2015 remained vague and did not establish any firm commitments to address climate change and mitigate greenhouse gas (GHG) emissions. It established that Qatar is 'extremely vulnerable to sea-level rise' and in particular addresses

(1) economic diversification with mitigation co-benefits (energy efficiency, clean energy and renewables, research and development, education, tourism), (2) adaptation actions with mitigation co-benefits (water management, infrastructure and transport, waste management, and awareness), and (3) response measures to climate change (Qatar 2015). Hence, there has not been a clear focus on mitigating climate change (but exports of LNG have been portrayed as contributing to emission reductions elsewhere). Qatar's first NDC did not include GHG emission targets nor did it identify sectors for mitigation action. Plans to invest in and use renewable energy sources, especially solar energy, were mentioned but, again, no concrete actions were taken or targets established. Instead, the NDC highlighted the need for technological assistance and technology transfers (Qatar 2015; see also Carbon Brief 2015b; WRI 2021). Overall, Qatar's first NDC reflects a lack of commitment to take ambitious action to mitigate climate change. As of July 2021, Qatar has not submitted any update of its first NDC, nor a second NDC.

According to the UNFCCC, Qatar's GHG emissions excluding land use, land use change, and forestry (LULUCF) in 2007 stood at 61,592.97 Gg CO_2 equivalent. Given the size and geographical location of Qatar, the LULUCF sector is insignificant (UNFCCC 2007). Qatar has the largest per capita emissions in the world (Qatar 2011). According to Qatar's National Communication to the UNFCCC, the oil and gas sector accounted for half of the country's GHG emissions, while the power and water sectors contributed 27 percent, road transport about 7 percent, and industrial processes 8.5 percent (Qatar 2011). The available UNFCCC data suggest over 90 percent of emissions are related to energy (UNFCCC 2007). Other more recent datasets suggest that emissions have increased more than seven-fold between 1990 and 2018 to nearly 100 Mt CO_2 equivalent (WRI 2021, see Figure 6.4).

Qatar does not have a firm framework for its domestic climate policy that is in embryonic state. In line with its NDC under the Paris Agreement, existing plans and measures

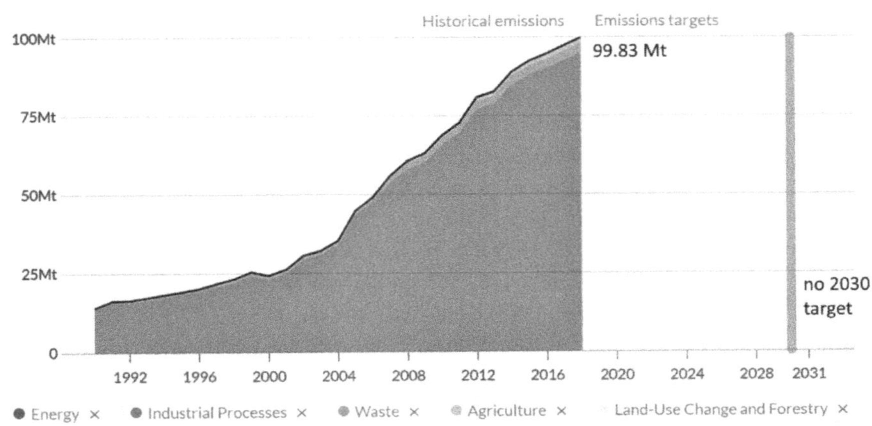

Figure 6.4 Qatar's greenhouse gas emissions and emission targets (1990–2018).
Source: WRI (2021).

seem to focus on the economy, with side or co-benefits for GHG mitigation. Instead of binding legislation, Qatar mainly works with non-binding initiatives, programmes, and incentives. Economic development is considered the primary concern, with climate change mitigation given a lower priority. Accordingly, the World Energy Council has ranked Qatar high on energy security and equity but low on environmental sustainability (where it, despite improvements over the 2010s, remains on one of the last ranks) in its Energy Trilemma Index (World Energy Council 2021). Although Qatar can also look back on a history of failed attempts to expand the use of renewable energy, there are signs that this may be emerging once again as a priority (see below).

The development of domestic climate policies is not helped by the close entanglement of Qatar's oil and gas sectors with the government. Qatar Petroleum, the dominant oil and gas company, is a state-owned public corporation which was established in 1974. It is responsible for the oil and gas sector in Qatar and is engaged in the exploration, production, and sale of, among others, crude oil, natural gas and gas liquids, refined products, and LNG in Qatar. The company's chairman is the Minister of Energy and Industry, but it also receives direct guidance from the Emir himself (Qatar Petroleum 2014, 2018). Qatar Petroleum operates several subsidiaries. One of these, QatarGas, is responsible for the development, production, and management of the country's gas fields and is the world's largest LNG producer (QatarGas 2018).

Trends and potential

Qatar has strong potential and has invested heavily in education and developing a knowledge-based economy. This forms part of a longstanding strategy of diversifying the economy beyond oil and gas production, including the expansion of renewable energy. While some progress has been made, related efforts are still at an early stage and leave significant room for further improvement.

Diversification of the economy

Also related to education and developing a knowledge-based economy are Qatar's longstanding efforts to diversify its economy, through both domestic and foreign investment (in particular through the Qatar Investment Authority). Progress has been made, but Qatar is pursuing further diversification in and away from the hydrocarbon sector in order to ensure a 'steady and robust' economy in the long term (Qatar 2011: 13; see also General Secretariat 2008; Qatar 2018). Past efforts have led to an expansion of gas production and exports (in large part via LNG capacities); development of petrochemical, metallic, and non-metallic industries; programmes for improving the quality of higher education; and expansion of the maritime and airline transport sectors. In addition, Qatar has allowed more privatisation in order to enable small- and medium-scale industries to grow faster. The construction sector (roads, high-rise buildings, private water and power projects, hospitality infrastructure, and residences) has been growing and making increasing contributions to GDP. Another growing sector is the transport and communications sector (Qatar 2011, 2018). Like small European countries, Qatar has faced challenges in the process of diversification such as a small domestic market, high production costs, low economies of scale, and low levels of industrial production (Miller and al-Mansouri 2016). Further diversification along similar lines is foreseen in Qatar's National Vision for 2030, the implementation of which is advanced through quintennial national development strategies (Qatar 2018). Particular emphasis has been put on developing a knowledge-based economy (see above), with a stronger focus on the private sector and more investment in infrastructure (General Secretariat 2008). Some advances towards a knowledge-based economy have been made, with some further room for improvement (Ben Hassen 2021).

However, from a decarbonisation perspective, it is worth highlighting that Qatar's diversification strategy does not appear to take into account the need to decarbonise and the risks associated with investments vulnerable to global decarbonisation. In other words, investments are not being vetted to assess whether they are climate-proof. Hence, as mentioned above, expansion of the gas sector has been part of the diversification strategy (to reduce dependence on oil). Qatar's National Vision 2030 considers the oil and gas sector to be of central importance for the development of human resources and economic capacities, and, in this context, gas is considered a source of clean energy (General Secretariat 2008: 32; Qatar 2018). Furthermore, efforts to build a domestic industrial base have been largely focused on the petrochemical industry, which itself is highly dependent on oil and gas and vulnerable to decarbonisation. Foreign industrial investments (e.g., in the European car industry) run related risks. Under the circumstances, the emphasis of Qatar's NDC on the possible negative impacts of climate change mitigation measures on the Qatari economy is understandable (Qatar 2015). However, Qatar lags in its own efforts to enhance its resilience.

Renewable energy and clean energy

As mentioned above, renewable energy seems to be emerging as a new priority field for action. This contrasts with a lack of delivery on past plans for renewable energy expansion. These included the target to generate 10 percent of total energy use for electricity and water desalination with solar power by 2018 (Arabian Business 2012). In 2019, renewable energy generation still amounted to a negligible 43 MW (of which 38 MW were bioenergy/waste and only 5 MW photovoltaics: IRENA 2021). In 2017, Qatar published a renewable energy strategy aimed at cutting emissions and diversifying the energy sector and the economy (Varghese 2017). Qatar has set a target to generate 2 percent of its electricity with renewable energy sources by 2020 and 20 percent by 2030 (Oxford Business Group 2017; Ren21 2018). It also intends to power the football stadiums built for the 2022 FIFA World Cup with solar power. However, 2020 data showed that renewable energy sources were, due to delays in planned investments, not yet contributing significantly to power generation (BP 2020). In 2020, Japan's Marubeni Corporation and French energy major Total won the tender to build an 800 MW solar power project, with a scheduled start of operation in 2022 (delayed from previous plans for several years; Renewables Now 2020). In short, much of the country's considerable potential for concentrated solar power, wind, and photovoltaics (Flamos 2015) has yet to be realised.

Education and knowledge-based economy

Qatar has a solid basic educational system, although there is still scope for further development and improvements. According to the World Bank's Human Capital Index, the level of education and healthcare would allow the average child born in Qatar in 2020 to achieve around 64 percent of their potential, which is above the average for the Middle East and North Africa, but below the average for high-income countries (World Bank 2020). In 2019, Qatari students were expected to attend school for 12 years (UNDP 2020). In the academic year 2014/2015, 77.8 percent of the relevant age groups attended secondary school and about 10 percent of them attended university (Qatar Ministry of Development Planning and Statistics 2017). Expenditure for education reached a peak of around 15 percent of total government expenditure in 2008 and accounted for 8.6 percent in 2019 (World Bank 2021g). Since 2010, expenditure on education per student has been among the highest in the world (De Bel-Air 2014). In the 2018 Program for International Student Assessment (PISA), Qatar was ranked 60 out of 77 countries (according to countries' performance in the reading category). While the country scored below the Organisation for Economic Co-operation and Development (OECD) average in all three main categories (reading, mathematics, and science), it performed on average in indicators related to socioeconomic background and obtained above average values on indicators related to immigration backgrounds (OECD 2019).

Qatar does have the ambition to develop a world-class educational system to underpin the development of a knowledge-based economy. To this end, the National Vision 2030 envisages the creation of a national network of formal and non-formal educational programmes (General Secretariat 2008). The country wants to 'accomplish a culture of innovation through investment in research and development' (Mohtar 2015: 49). To realise this ambition, the Qatari government has announced it will invest 2.8 percent of its GDP (US$2–3 billion) into research and development (General Secretariat 2009: 24; see also Qatar 2018). The Qatar National Research Fund that issues grants for different research programmes and the Qatar Science Leadership Programme that aims to motivate students to pursue careers in science and research are important tools in this context.

The significant expatriate workforce is also an important factor to consider in this respect. Currently, this workforce is mainly low skilled. Qatar's National Vision for 2030 therefore aims to improve 'the size and quality of the expatriate labour force and the selected path of development' in particular by promoting a recruitment policy 'in line with a knowledge-based economy' (De Bel-Air 2014: 5). Efforts aim to balance 'local and imported talent' (Mohtar 2015: 54).

EU-Qatar cooperation

EU-Qatar relations have grown closer over the 2010s, and a Cooperation Agreement was signed in 2018. Furthermore, EU-Qatar relations are embedded in broader EU-GCC cooperation. Trade and investment as well as energy are important and evolving areas, with significant potential to integrate decarbonisation. Given the precarious situation in the Middle East, security and geopolitics form another key area of the relations between Qatar and the EU, with EU member states playing a central role.

Institutional framework: Cooperation Agreement and embedding in the Gulf Cooperation Council

Relations between the EU and Qatar have grown closer in the 2010s, particularly in certain economic areas. Since 2011, there have been annual rounds of high-level strategic dialogue between Qatar and the EU (Miller and al-Mansouri 2016), and a dedicated GCC-EU dialogue on trade and investment issues was launched in 2017 (European Commission 2018a). In March 2018, a Cooperation Agreement between the EU and Qatar was signed as a basis for 'enhanced political dialogue and strengthened cooperation on sectoral areas of mutual interests' (EEAS 2018), such as private sector development and research and innovation. These processes are intended to support Qatar in realising its National Vision for 2030. Qatar also maintains important bilateral relations with several EU member states, such as France, Germany, Italy, Spain, and some of the smaller EU member states (as well as the United Kingdom) (Miller and al-Mansouri 2016).

Relations between the EU and Qatar have been very much embedded in the broader relations between the EU and the GCC. Founded in 1981, the GCC consists of Bahrain, Kuwait, Oman, Qatar, Saudi Arabia, and the United Arab Emirates. An EU-GCC Cooperation Agreement was adopted in 1989. A Joint Council and a Joint Cooperation Committee under the Agreement both meet annually. Cooperation covers trade and investment issues, macroeconomic matters, climate change, energy, environment, and research. Negotiations on a Free Trade Agreement were launched in 1990, but GCC countries suspended them in 2008. As mentioned above, since 2010 increasing tensions between GCC countries and especially between Qatar and Saudi Arabia have called into question the future role of the GCC (including for cooperation with the EU). Qatar's announcement in December 2018 to leave OPEC points in a similar direction. It remains to be seen to what extent the settlement reached between Qatar and Saudi Arabia within the GCC in January 2021 will result in a lasting solution.

Trade and investment

There is already significant trade and investment between Qatar and the EU, and there is potential for increasing both in future. The EU's total trade in goods with Qatar in 2019 (i.e., pre-Covid-19) was around €17 billion, with the EU exporting about €11.4 billion to Qatar and importing around €5.8 billion from Qatar (European Commission 2021). Nearly half of EU exports to Qatar were machinery, while gas and oil accounted for more than two-thirds of imports from Qatar. As a result of the Covid-19 crisis, EU-Qatar trade (both imports and exports) contracted by more than a third in 2020 (European Commission 2021). In 2019, around 4 percent of EU import of natural gas came from Qatar (Chatham House 2021). EU exports accounted for more than 35 percent of Qatari imports, whereas EU imports made up little more than 10 percent of Qatar's exports. Trade in services was somewhat less significant at around €6 billion (2016), accounting for close to 15 percent of Qatar's total trade in services (European Commission 2018c; see also Miller and al-Mansouri 2016). Like all GCC countries, Qatar is classified as a high-income country by the World Bank and therefore does not benefit from the EU Generalized Scheme of Preferences (European Commission 2018a). It is also not a recipient of EU aid (European Commission 2018b).

Trade in fossil fuels is relatively modest for both sides. In 2019, Japan and the Republic of Korea were the top importers of Qatar's fossil fuels, with shares of 19 and 17 percent, respectively. India accounted for a share of 13 percent, and China 12.5 percent. As visible from the above trade figures, the EU received about 8 percent of Qatari fossil fuel exports, with Italy (2 percent) and Spain (1.4 percent) as important destinations (Chatham House 2021). Nevertheless, Qatar is an important partner for the EU's efforts to diversify gas supplies (see also Figure 6.1).

The investment relationship between Qatar and the EU is also significant and inverse to trade in comparison. The stock of EU foreign direct investments (FDI) in Qatar amounted to €6 billion in 2016, whereas Qatari FDI in the EU stood at €20.8 billion (European Commission 2018c). This reflects that Qatar considered the economic crisis that struck Europe after 2008 an investment opportunity. Qatar used the situation to buy European sovereign debt and rescued several financial institutions (Miller and al-Mansouri 2016). Qatar has made various strategic investments in EU companies (e.g., Siemens and Volkswagen). The EU and Europe are therefore an important partner for Qatar in its efforts to diversify its economy by investing in other economic activities.

Energy

Beyond existing relations on trade and investment, in particular regarding oil and gas (see above), the potential for cooperation on (renewable and clean) energy remains to be exploited to a large extent. The EU GCC Clean Energy (Technology) Network has the ambition to catalyse and coordinate joint EU-GCC clean energy initiatives, including policy and technology aspects. The network's main instruments are activities such as policy discussions, joint research, or technology implementation projects organised in a system of different Working Groups and a web-based cooperation platform (European Commission 2016). However, it has not yet led to a measurable increase in renewable and clean energy in Qatar (BP 2020; IRENA 2021).

Carbon Capture and Storage (CCS) has been identified as another potential area for cooperation. CCS has the potential to contribute to emission reductions in the power and industry sectors. CCS may be attractive for Qatar and other GCC states because it could 'green' natural gas-fired power generation and emissions-intensive industrial processes such as LNG production, as well as being used for enhanced recovery of oil and natural gas (Meltzer et al. 2014; Flamos 2015). Its use in enhanced oil recovery could also reduce the cost of oil production. Qatar has itself developed some CCS projects in recent years, some in cooperation with international companies such as Shell, and others based at the Qatari University. Some GCC countries (Qatar, Saudi Arabia, and Kuwait) have also cooperated on funding CCS research (Meltzer et al. 2014; Al-Yaeeshi et al. 2018; Adler 2021).

Clean energy cooperation between the EU and GCC has been challenging due to GCC and EU countries' different approaches to climate change. While the EU has developed a broad policy framework with legislation and other measures to mitigate GHG emissions, GCC countries (including Qatar) are more concerned about the effects of climate policy on their economies (Flamos 2015).

Security and geopolitics

The geopolitical and security component of EU-Qatar relations also deserves highlighting. The EU and its member states are only some among the many players in the delicate regional security balance between Saudi Arabia and Iran with the involvement of other regional players, including the United States and Russia (but also Great Britain). These security and defence aspects are particularly relevant to Qatar's bilateral relations with France (which is, together with Great Britain, a significant arms supplier to Qatar), while the United States is the dominant external regional security actor and a crucial partner for Qatar (Miller and al-Mansouri 2016). While socioeconomic aspects remain most prominent in relations between the EU and Qatar, the EU has offered to support Kuwait in mediating in the GCC crisis (EEAS 2018). Overall, the role of the EU has been rather that of a mediating power. While this has had its value for Qatar, the EU did not play a significant role in ending the embargo by Saudi Arabia and others, as the related agreement was brokered by Kuwait with support from the United States (see above).

Conclusions

Qatar is highly dependent on oil and gas exports for its economic welfare and political stability. Hence, decarbonisation in Europe and the world has the potential to undermine the political and economic stability of Qatar. Phasing out fossil fuel production and exports will mean replacing Qatar's main engine of economic development and source of government income, which could reinforce existing risks to political stability, including limited political freedom and the integration of a large number of (low-skilled) expatriate workers. Decarbonisation hence poses the challenge to develop alternatives to oil and gas as a basis of Qatar's economic well-being and political settlement. Gas likely remaining a significant fuel in the transition for some time provides the opportunity the develop alternatives in a stepwise process.

Risks to political stability in Qatar could have important wider regional ramifications. Being located in the Gulf region, Qatar is part of a politically delicate regional power play and balance with a high potential for conflict – as is evident from the blockade against Qatar by several of its regional neighbours led by Saudi Arabia from 2017 to 2021. Also, neighbouring countries face very similar challenges under decarbonisation. Hence, decarbonisation requires proactive transition management in order to contain stability risks across the region more broadly. Such transition management may best address relations with Qatar in the context of an integrated approach towards the Gulf region and the Gulf Cooperation Council as its main regional organisation.

There are a number of entry points for fruitfully developing the relationship between Qatar (and the Gulf region) and the EU under decarbonisation. At a geopolitical level, the EU has an interest in developing this relationship because of the central role of the Middle East in its neighbourhood and the potential for diversifying its relations with the region (and strengthening its role as a neutral bridge builder). On its side, Qatar – given continuing intra-regional tensions – has a growing interest in diversifying its external relations and opportunities. More specifically related to the decarbonisation challenge, Qatar has an interest in developing its economy beyond oil (as evidenced by its recent withdrawal from OPEC) and eventually also beyond gas (which, however, will remain of strategic importance in the medium term). The EU can offer to be a partner in important ways to this end. The rather limited fossil fuel/oil relationship between the EU and Qatar may provide a useful background in this respect.

On the basis of the preceding analysis, we can identify a number of prominent and promising focal areas for developing the relationship between the EU and Qatar beyond oil and gas:

- **Education and training** are a central part of the effort at economic diversification and developing a knowledge-based economy and a priority for Qatar. Bilateral cooperation to this end could be developed to bring the significant resources of the EU and its member states to bear to assist in enhancing and shaping education and training in Qatar (and the wider Gulf region). One significant focus in this respect could be cooperation supporting the expansion of renewable energy. Concretely, partnerships between universities in the EU and Qatar could be fostered.
- **Further developing the energy system, and particularly expanding renewable energy**, could result in numerous benefits by reducing economic dependence on oil and gas. As well as addressing the urgent need to decarbonise, it also makes sense in terms of Qatar's continued exploitation of its oil and gas resources. Increasing the proportion of renewables in domestic power generation would increase the share of resources available for export. Concrete first steps may include cooperation in the installation of renewable energy capacity for the 2022 FIFA World Cup (and beyond) and creation of a policy dialogue on the potential of clean/renewable energy (possibly in a GCC context). Such efforts could also be developed into a more comprehensive partnership for climate and clean energy which could integrate several elements, including development of the adaptation-mitigation nexus (e.g., benefits of renewables development in building resilience), a medium-term strategy for the development of the gas sector as a bridge to decarbonisation, and possibly the development of renewables-based hydrogen production.
- **Advancing mutual market access and investments**. Qatar's efforts to develop its external investment portfolio and to spur foreign direct investment in Qatar (as part of its diversification strategy) provide a fruitful basis for further advancing economic cooperation. The EU and its member states possess significant experience and expertise in this regard and are attractive for and engage in FDI. Part of enhancing cooperation in this area could be a dialogue on strengthening financial governance, including the Qatar Investment Authority and the changing landscape of climate-proofing investments. Again, renewable and clean energy may be an obvious focal area.
- Multiple other areas can be further developed over time, including water management and security cooperation.

Overall, there is a range of options for developing the EU-Qatar relationship under decarbonisation. Such a renewed direction for the bilateral relationship would promise significant benefits for Qatar as it would support the country in advancing its own diversification strategy and making it climate-proof. This may in turn increase awareness in Qatar about the needs, feasibility, and benefits of moving beyond oil and gas. At the same time, it would help make EU-Qatar relations fit for a decarbonising world and may make a significant contribution to stabilising the geopolitically precarious Gulf region in the context of decarbonisation.

Note

1 Production, consumption, and export data vary between different sources (including IEA, CIA, BP Statistical Review) but all give roughly the same overall picture.

References

Adler, K. (2021) 'Qatar Petroleum commits to low-carbon LNG in latest expansion', IHS Markit. 11 February 2021. Retrieved 12 July 2021 from https://ihsmarkit.com/research-analysis/qatar-petroleum-commits-to-lowcarbon-lng-in-latest-expansion.html.

Al-Sarihi, A. (2018) *Prospects for Climate Change Integration Into the GCC Economic Diversification Strategies*. LSE Middle East Centre Paper Series, 20 February 2018. London. Available at: http://eprints.lse.ac.uk/86873/1/Al-Sarihi_Prospects%20for%20climate%20change_2018.pdf (accessed 21 April 2020).

Al-Yaeeshi, A.A., Al-Ansari, T., and Govindan, R. (2018) 'The potential for carbon dioxide capture and utilization within the State of Qatar', *Computer Aided Chemical Engineering*, 43, pp. 1499–1504.

Amnesty International (2020) *Reality Check 2020: Countdown to the 2022 World Cup*. Available at: https://www.amnesty.org/download/Documents/MDE2232972020ENGLISH.PDF (accessed 9 July 2021).

Arabian Business (2012) 'Qatar eyes 10% of energy from renewable sources', 22 May 2012. Retrieved 21 April 2020 from https://www.arabianbusiness.com/qatar-eyes-10-of-energy-from-renewable-sources-458792.html.

Ben Hassen, T. (2021) 'The state of the knowledge-based economy in the Arab world: Cases of Qatar and Lebanon', *EuroMed Journal of Business*, 16(2), pp. 129–153. https://www.emerald.com/insight/content/doi/10.1108/EMJB-03-2020-0026/full/html.

Bertelsmann Stiftung (2020) *BTI 2020 Country Report. Qatar*. Gütersloh: Bertelsmann Stiftung. Available at: https://www.bti-project.org/content/en/downloads/reports/country_report_2020_QAT.pdf (accessed 14 June 2021).

BP (2020) *BP Statistical Review of World Energy*. Available at: https://www.bp.com/content/dam/bp/business-sites/en/global/corporate/pdfs/energy-economics/statistical-review/bp-stats-review-2020-full-report.pdf (accessed 9 June 2021).

Carbon Brief (2015a) 'Paris Climate Talks: Who Are the Negotiating Groups?' Retrieved 21 April 2020 from https://www.carbonbrief.org/interactive-the-negotiating-alliances-at-the-paris-climate-conference.

Carbon Brief (2015b) 'Paris 2015: Tracking Country Climate Pledges'. Retrieved 21 April 2020 from https://www.carbonbrief.org/paris-2015-tracking-country-climate-pledges.

Chatham House (2021) 'Resources Trade. Earth'. Retrieved 14 June 2021 from https://resourcetrade.earth.

CIA (2021) 'The World Factbook. Qatar'. Retrieved 15 June 2021 from https://www.cia.gov/the-world-factbook/countries/qatar/.

Coates Ulrichsen, K. (2010) *The GCC States and the Shifting Balance of Global Power*. Prepared on behalf of the Centre for International and regional Studies. Georgetown University. Available at: https://www.files.ethz.ch/isn/124456/No_6_KristianCoatesUlrichsenOccasionalPaper.pdf (accessed 21 April 2020).

Collins, G. (2018) *Anti-Qatar Embargo Grinds Toward Strategic Failure.* Issue Brief, 1. Rice University's Baker Institute for Public Policy. Available at: https://scholarship.rice.edu/bitstream/handle/1911/99690/bi-brief-012218-ces-qatarembargo.pdf?sequence=1&isAllowed=y (accessed 9 July 2021).

De Bel-Air, F. (2014) *Demography, Migration, and Labour Market in Qatar. Gulf Labour Markets and Migration*. GLMM No. 8/2014. Available at: https://cadmus.eui.eu/bitstream/handle/1814/32431/GLMM_ExpNote_08-2014.pdf?sequence=1&isAllowed=y (accessed 9 July 2021).

EEAS – European External Action Service (2018) 'EU and Qatar Sign a Cooperation Agreement'. Retrieved 21 April 2020 from https://eeas.europa.eu/headquarters/headquarters-Homepage/40967/eu-and-qatar-sign-cooperation-arrangement_en.

EIA – Energy Information Agency (2015) 'Qatar'. Retrieved 21 April 2020 from https://www.eia.gov/beta/international/analysis.php?iso=QAT.

European Commission (2016) *O.1.6. Report. Background Paper on Areas of Potential EU GCC Clean Energy Cooperation.* Available at: https://eugcc-cleanergy.net/sites/default/files/o_1_6_background_paper_cenii_v13_07102016_-clean.pdf (accessed 21 April 2020).

European Commission (2018a) 'Gulf Region'. Retrieved 21 April 2020 from http://ec.europa.eu/trade/policy/countries-and-regions/regions/gulf-region/.

European Commission (2018b) 'EU Aid Explorer'. Retrieved 21 April 2020 from https://euaidexplorer.ec.europa.eu/AidOverview.do.

European Commission (2018c) 'Qatar'. Retrieved 21 November 2019 from http://trade.ec.europa.eu/doclib/docs/2006/september/tradoc_111844.pdf.

European Commission (2021) 'European Union Trade in Goods with Qatar'. Retrieved 14 July 2021 from https://webgate.ec.europa.eu/isdb_results/factsheets/country/details_qatar_en.pdf.

Federal Reserve Bank of St. Louis (2018) 'Non-oil Revenue for General Government for Qatar'. Retrieved 21 April 2020 from https://fred.stlouisfed.org/series/QATGGRXOGDPXOPT.

Flamos, A. (2015) 'The timing is ripe for an EU-GCC "Clean Energy" Network', *Energy Sources, Part B: Economics, Planning, and Policy*, 10(3), pp. 314–321.

Freedom House (2021) 'Freedom in the World 2021. Qatar Profile'. Retrieved 14 June 2021 from https://freedomhouse.org/country/qatar/freedom-world/2021.

Fund for Peace (2021) 'Fragile States Index 2021'. Retrieved 26 July 2021 from https://fragilestatesindex.org/.

Galeeva, D. (2018) 'How National Identity Will Shape the Future of Liberalism: The Consequences of Brexit in the EU, and of the 2017 Crisis in the GCC'. Retrieved 21 April 2020 from https://mesbar.org/national-identity-will-shape-future-liberalism-consequences-brexit-eu-2017-crisis-gcc/.

General Secretariat for Development Planning (2008) *Qatar National Vision 2030*. Available at: https://www.psa.gov.qa/en/qnv1/Documents/QNV2030_English_v2.pdf (accessed 21 April 2020).

General Secretariat for Development Planning (2009) *Advancing Sustainable Development. Qatar National Vision 2030.* Available at: http://hdr.undp.org/sites/default/files/qhdr_en_2009.pdf (accessed 9 October 2019).

Germanwatch (2021) 'Global Climate Risk Index 2021'. Retrieved 14 June 2021 from https://germanwatch.org/de/19777.

Hydrocarbons Technology (2018) 'Barzan Gas Project, Ras Laffan Industrial City, Qatar'. Retrieved 21 April 2020 from https://www.hydrocarbons-technology.com/projects/barzan-gas-project-ras-laffan-qatar/.

IEA – International Energy Agency (2015) 'Coal for 2015'. Retrieved 15 November 2019 from https://www.iea.org/statistics/statisticssearch/report/?year=2015&country=Qatar&product=Coal.

IEA (2020) 'Energy Subsidies'. Retrieved 2 July 2021 from https://www.iea.org/weo/energysubsidies/.

IRENA – International Renewable Energy Agency (2021) 'Energy Profile Qatar'. Retrieved 14 June 2021 from https://www.irena.org/IRENADocuments/Statistical_Profiles/Middle%20 East/Qatar_Middle%20East_RE_SP.pdf.

Meltzer, J., Hultman, N., and Langley, C. (2014) *Low-Carbon Energy Transitions in Qatar and the Gulf Cooperation Council Region*. Brookings. Available at: https://www.brookings.edu/wp-content/uploads/2016/07/low-carbon-energy-transitions-qatar-meltzer-hultman-full.pdf (accessed 21 April 2020).

Miller, R. and al-Mansouri, K. (2016) 'Qatar's foreign policy engagement with the European Union: Evolving priorities of a small state in the contemporary era', *Comillas Journal of International Relations*, 5, pp. 43–64. https://doi.org/10.14422/cir.i05.y2016.004.

Mohtar, R.H. (2015) 'Opportunities and challenges for innovations in Qatar', *The Muslim World*, 105(1), pp. 46–57. https://doi.org/10.1111/muwo.12079.

ND-GAIN – Notre Dame Global Adaptation Initiative (2020) 'ND-GAIN Country Index'. South Bend, IN: University of Notre Dame. Retrieved 5 July 2021 from https://gain.nd.edu/our-work/country-index/.

OECD – Organisation for Economic Co-operation and Development (2019) *PISA 2018 Results. Combined Executive Summaries Volume I, II & III*. Available at: https://www.oecd.org/pisa/Combined_Executive_Summaries_PISA_2018.pdf (accessed 14 June 2021).

Offshore Energy Today (2017) 'Qatar developing new gas project in giant offshore field', 3 April 2017. Retrieved 21 April 2020 from https://www.offshoreenergytoday.com/qatar-petroleum-developing-new-offshore-gas-project/.

Oxford Business Group (2017) 'Qatar gets serious about solar', 21 April 2017. Retrieved 21 April 2020 from https://oxfordbusinessgroup.com/news/qatar-gets-serious-about-solar.

Qatar (2011) *Initial National Communication to the United Nations Framework Convention on Climate Change*. Qatar Ministry for the Environment. Retrieved 21 April 2020 from http://unfccc.int/files/national_reports/non-annex_i_natcom/application/pdf/final_climate_change.pdf.

Qatar (2015) *Intended Nationally Determined Contributions (INDC) Report*. Retrieved from https://www4.unfccc.int/sites/ndcstaging/PublishedDocuments/Qatar%20First/Qatar%20INDCs%20Report%20-English.pdf.

Qatar (2018) *Qatar Second National Development Strategy 2018–2022*. Doha: Planning and Statistics Authority. Available at: https://www.psa.gov.qa/en/knowledge/Documents/NDS2Final.pdf (accessed 6 July 2021).

QatarGas (2018) 'The World's Premier LNG Company'. Retrieved 21 April 2020 from http://www.qatargas.com/english/aboutus/.

Qatar Investment Authority (2016) 'Building the Future with Global Vision'. Retrieved 21 April 2020 from http://www.qia.qa/.

Qatar Ministry of Development Planning and Statistics (2017) *Education in Qatar. Statistical Profile 2016*. Available at: https://www.mdps.gov.qa/en/statistics/Statistical%20Releases/Social/Education/2016/Education_Statistical_Pro%EF%AC%81le_2016_En.pdf (accessed 21 April 2020).

Qatar Petroleum (2014) *1974–2014 Years of Excellence. Annual Report 2014*. Available at: https://www.qp.com.qa/en/MediaCentre/Documents/2014%20Annual%20Report%20-%20English.pdf (accessed 21 April 2020).

Qatar Petroleum (2018) 'About QP'. Retrieved 21 April 2020 from https://www.qp.com.qa/en/AboutQP/Pages/AboutUs.aspx.

Quandl (2021) 'Qatar General Government Revenues, USD Billions'. Retrieved 14 June 2021 from https://www.quandl.com/data/ODA/QAT_GGR-Qatar-General-Government-Revenue-USD-Billions.

REN21 – Renewable Energy Policy Network for the 21st Century (2018) *Renewables 2018. Global Status Report*. Available at: https://www.ren21.net/wp-content/uploads/2019/05/GSR2018_Full-Report_English.pdf (accessed 21 April 2020).

Renewables Now (2020) 'Update – Marubeni, total to build 800-MWp solar park in Qatar', 20 January 2020. Retrieved 4 July 2021 from https://renewablesnow.com/news/update-marubeni-total-to-build-800-mwp-solar-park-in-qatar-684035/.

Reiche, D. (2010) 'Energy policies of Gulf Cooperation Council (GCC) countries – possibilities and limitations of ecological modernization in rentier states', *Energy Policy*, 38(5), pp. 2395–2403.

Salam, M. and Alam, H.A. (2021) 'Saudi Arabia and Qatar agree to reopen airspace and maritime borders', *CNN*, 5 January 2021. Retrieved 9 July 2021 from https://edition.cnn.com/2021/01/04/world/qatar-and-saudi-arabia-reopen-airspace-intl/index.html.

Trade Arabia (2018) 'Qatar set to Award $85 bn Construction Projects', 1 August 2018. Retrieved 21 April 2020 from http://www.tradearabia.com/news/OGN_343571.html.

Trading Economics (2021) 'Qatar Government Debt to GDP'. Retrieved 2 July 2021 from https://tradingeconomics.com/qatar/government-debt-to-gdp?continent=g20.

Transparency International (2021) 'Corruption Perceptions Index'. Retrieved 14.06.2021 from https://www.transparency.org/en/cpi/2020/index/qat.

UN DESA – United Nations Department of Economic and Social Affairs (2019). 'World Population Prospects 2019. Total Population by Sex (Thousands). Qatar'. Retrieved 9 June 2021 from https://population.un.org/wpp/DataQuery/.

UNDP – United Nations Development Programme (2020) 'Human Development Report'. Available at: http://hdr.undp.org/en/countries (accessed 26 July 2021).

UNFCCC – United Nations Framework Convention on Climate Change (2007) 'Emissions Summary for Qatar'. Retrieved 21 April 2020 from https://unfccc.int/files/ghg_data/ghg_data_unfccc/ghg_profiles/application/pdf/qat_ghg_profile.pdf.

Varghese, J. (2017) 'Qatar's first renewable energy strategy on the anvil', *Gulf Times*, 26 November 2017. Retrieved 21 April 2020 from http://www.gulf-times.com/story/572622/Qatar-s-first-renewable-energy-strategy-on-the-anv.

World Bank (2020) '2020 Human Capital Index: Country Briefs and Data'. Retrieved 1 July 2021 from https://www.worldbank.org/en/publication/human-capital#Index.

World Bank (2021a) 'GDP (Current US$)'. Retrieved 9 June 2021 from https://data.worldbank.org/indicator/NY.GDP.MKTP.CD?locations=QA.

World Bank (2021b) 'GDP Per Capita (Current US$)'. Retrieved 9 June 2021 from https://data.worldbank.org/indicator/ny.gdp.pcap.cd?locations=QA.

World Bank (2021c) 'GDP (Constant 2010 US$)'. Retrieved 11 June 2021 from https://data.worldbank.org/indicator/NY.GDP.MKTP.KD?locations=QA.

World Bank (2021d) 'GDP Per Capita (Constant 2010 US$)'. Retrieved 11 June 2021 from https://data.worldbank.org/indicator/NY.GDP.PCAP.KD?locations=QA.

World Bank (2021e) 'Oil Rents as Percent of GDP, 2018'. Retrieved 9 June 2021 from https://data.worldbank.org/indicator/NY.GDP.PETR.RT.ZS.

World Bank (2021f) 'Gas Rents as Percent of GDP, 2018'. Retrieved 9 June 2021 from https://data.worldbank.org/indicator/NY.GDP.NGAS.RT.ZS.

World Bank (2021g) 'Government Expenditure on Education, Total (% of Government Expenditure)'. Retrieved 14 June 2021 from https://data.worldbank.org/indicator/SE.XPD.TOTL.GB.ZS?locations=QA&most_recent_value_desc=true.

World Bank (2021h) 'Qatar'. Retrieved 6 July 2021 from https://thedocs.worldbank.org/en/doc/6853dd5c93a231b05c90c7a1fd7ac5b4-0280012021/original/14-mpo-sm21-qatar-qat-kcm4.pdf.

World Bank (2021i) 'Fuel Exports (% of Merchandise Exports)'. Retrieved 15 July 2021 from https://data.worldbank.org/indicator/TX.VAL.FUEL.ZS.UN?locations=QA.

World Bank (2021j) 'Worldwide Governance Indicators 2019'. Retrieved 26 July 2021 from http://info.worldbank.org/governance/wgi/#reports.

World Bank (2021k) *Regulatory Indicators for Sustainable Energy (RISE)*. Retrieved 26 July 2021 from https://rise.esmap.org.

World Energy Council (2021) 'Energy Trilemma Index'. Retrieved 14 June 2021 from https://trilemma.worldenergy.org/#!/country-profile?country=Qatar&year=2020.

WRI – World Resources Institute (2021) 'Climate Watch. Qatar: Greenhouse Gas Emissions and Emissions Targets'. Retrieved 17 June 2021 from https://www.climatewatchdata.org/countries/QAT.

Zhdannikov, D. (2018) 'Qatar seems stampede for gas projects to help beat crisis', *Reuters*, 24 January 2018. Retrieved 21 April 2020 from https://www.reuters.com/article/us-davos-meeting-qatar-energy/qatar-sees-stampede-for-gas-projects-to-help-beat-crisis-idUSKBN1FD0R6.

7 Canada

Reframing a well-established partnership

Introduction

Canada is a country of more than 37 million people (UN DESA 2019), with Ottawa as its capital. It is the world's second largest country by area (including waters) and stretches from the Atlantic Ocean in the east to the Pacific Ocean in the west, with the Arctic Ocean to the north. Its neighbours include the United States (with borders to the United States to the south and to the US state of Alaska to the northwest), Greenland to the northeast, and the French overseas territories of Saint Pierre and Miquelon to the southeast. Canada is an industrialised economy and a highly urbanised country with 83 percent of its population living in metropolitan areas and census agglomerations (Statistics Canada 2016), principally near its southern border with the United States. Canada has the tenth largest gross domestic product (GDP) in the world, amounting to US$1.736 trillion in 2019 (World Bank 2021b). Canada's debt-to-GDP ratio stood at 86.8 percent in 2019, down from 91.8 percent in 2016 (Trading Economics 2021). It is a stable federal parliamentary democracy and a constitutional monarchy, with the United Kingdom's Queen Elizabeth II as the head of state. It is one of the world's most ethnically diverse and multicultural nations.

The Canadian federal political system deserves particular attention. Canada is composed of ten provinces (Alberta, British Columbia, Manitoba, New Brunswick, Newfoundland and Labrador, Nova Scotia, Ontario, Prince Edward Island, Quebec, and Saskatchewan) and three territories (Northwest Territories, Nunavut, and the Yukon). Territorial governments exercise authority delegated by the federal government, whereas each of the ten provinces possesses a formally independent government (so that Canada is governed by 11 governments in total). Canadian federalism has its roots in longstanding differences and tensions, especially between French-speaking and English-speaking parts (with French-speaking Quebec hosting a strong independence movement). The provinces possess far-reaching authority independent of the federal government, including for the exploration, development, and export to other provinces of non-renewable natural resources, forestry resources, and electricity. Important aspects of climate and energy policy are therefore either partly or entirely determined at the provincial level. The division of powers between the provincial and federal levels has evolved over the years and has remained the subject of intense political and legal battles.

DOI: 10.4324/9781003183037-7

Among developed nations, Canada's economy is unusual given the relative importance of the primary and energy sector – logging, mining, and the oil industry are among the most important. Canada is also one of the few advanced economies to be a net exporter of energy. In 2019, Canada's energy sector constituted more than 10 percent of its GDP and 23 percent of its total goods exports, employing approximately 280,000 people or 1.5 percent of total employment (Government of Canada 2020a). Government revenues collected annually from the oil and gas industries averaged CAN$14 billion between 2015 and 2020 including CAN$11 billion from upstream oil and gas extraction and its support activities (Natural Resources Canada 2020). In addition, the Canadian forest sector provides employment to about 205,000 workers and accounts for about 4.3 percent of total Canadian exports, adding CAN$23.7 billion to the economy in 2019 (Government of Canada 2020b; World Bank 2021a). Canada also has a very significant mining sector beyond fossil fuels (including chromite and iron mining and others). It possesses about 8 percent of the world's unmined uranium resources and accounts for a quarter of global primary uranium production (Natural Resources Canada 2019).

Canada is highly interdependent with and dependent on the United States. The United States is the only country with which Canada has a (very long) land border, and the US and Canadian economies are closely intertwined. Economic and trade relations have further intensified through their embedding in the North American Free Trade Agreement (NAFTA) and its successor, the US-Mexico-Canada Agreement (in force since 2020). As a result, the United States is by far Canada's most important trading partner, accounting for about half of Canadian imports and three quarters of Canadian exports (European Commission 2021a). More than 90 percent of Canadian exports of oil and gas have the United States as their destination, mainly through several pipelines (also due to a lack of sufficient other export capacity in Canada) (Chatham House 2021).

Like the other case studies, this chapter will explore the challenges and opportunities Canada and Canada-EU relations face under decarbonisation in five steps. The next section first demonstrates Canada's dependence on the production and export of oil and gas and hence its vulnerability to price fluctuations and the phase-out of fossil fuels. It also investigates other fragility and security risks that are comparatively minor. Subsequently, we assess the status of Canada's climate policy. This is followed by the review of relevant trends and potentials, including in the services sector and renewable and clean energy. The one but last section reviews the broad framework and basis of EU-Canada relations. The final section concludes and identifies focal points for EU foreign policy towards Canada under decarbonisation.

Exposure and risk

Exposure to global decarbonisation trends

Canada is a major producer of oil, gas, and also coal. In 2019, it produced more than 5.6 million barrels of oil per day, equivalent to almost 6 percent of the world total and an increase of about 70 percent from 2009 levels. Canada's gas production exceeded 170 billion cubic metres in 2019 (up from about 150 billion cubic metres in 2012), which represented 4.3 percent of global production. Coal production in 2019 amounted to about 50 million tonnes, down from 68.4 million tonnes in 2013 and accounting for less than 1 percent of global production (BP 2020; see also EIA 2019; CIA 2021). The production of fossil fuels plunged in 2020 in the wake of the Covid-19 crisis.

Canada also possesses significant fossil fuel reserves, especially oil and gas. Its proven oil reserves – primarily in the form of oil sands – are estimated to amount to nearly 170 billion barrels, which could support current production levels for another more than 80 years. They are the third largest in the world (after Venezuela and Saudi Arabia) and account for 10 percent of the global total. Estimated gas reserves are less abundant and amount to 2 trillion cubic metres (BP 2020; see for other estimates: EIA 2019; CIA 2021). They could support current production levels only for another ten years. Canada also has significant reserves of shale gas that remain to be further explored. Proved coal reserves of about 6,600 million tonnes could support 2019 production levels for another 130 years (BP 2020; see also EIA 2019).

In accordance with its resource base, gas and oil supply a large part of Canada's energy needs. Oil consumption exceeded 2.4 million barrels per day in 2019, equivalent to about 45 percent of production. Gas consumption amounted to about 120 billion cubic metres in 2019, up from about 90 billion cubic metres in 2009 and equivalent to about 70 percent of production. Coal consumption reached 13.4 million tonnes oil equivalent, down from almost 23.4 million tonnes oil equivalent in 2009 and equivalent to about 50 percent of production. Overall, about 62 percent of overall energy consumption is accounted for by oil and gas, about 4 percent by coal, about 24 percent by hydroelectricity, and the remainder by nuclear power and renewable energy other than hydro (BP 2020).[1] Hydropower is the backbone of the electricity system, supplying around 60 percent of electricity consumption. Among the International Energy Agency (IEA) member countries, Canada has the highest energy supply per capita: total primary energy supply per capita in Canada in 2019 was 8.0 tonnes of oil equivalent versus an OECD (Organisation for Economic Co-operation and Development) average of 4.1 (IEA 2021).

Fossil fuel production and reserves display enormous regional differences across Canada. Gas, oil, and coal production and reserves are concentrated in the broader Western Canada Sedimentary Basin (in addition to oil production in the offshore oil fields in the Atlantic Ocean, which has been declining) and hence in Alberta (oil, gas, and coal), British Columbia (gas and coal), and Saskatchewan (oil and gas). By far the largest share of Canada's proved oil (oil sands) and gas reserves are located in Alberta. Consequently, Alberta accounts for more than three quarters of Canadian oil and gas production.

Energy exports are of great importance for Canada's economy. In 2019, the country exported more than 50 percent of its oil production, about 30 percent of its gas production, and about half of its coal production. Shares of actual exports of production are even higher, as Canada also imports these fossil fuels (BP 2020). Energy exports accounted for nearly a quarter of Canadian goods exports in 2019, with this share having fluctuated roughly between 20 and 30 percent since 2005, also as a result of price fluctuations (Government of Canada 2020a; World Bank 2021e; see Figure 7.1).

The United States is Canada's top energy trade partner. In 2019, the United States accounted for 90 percent (CAN$ 121.5 billion by value) of Canada's exported energy products: 96 percent of Canadian oil and gas exports totalling over CAN$117 billion went to

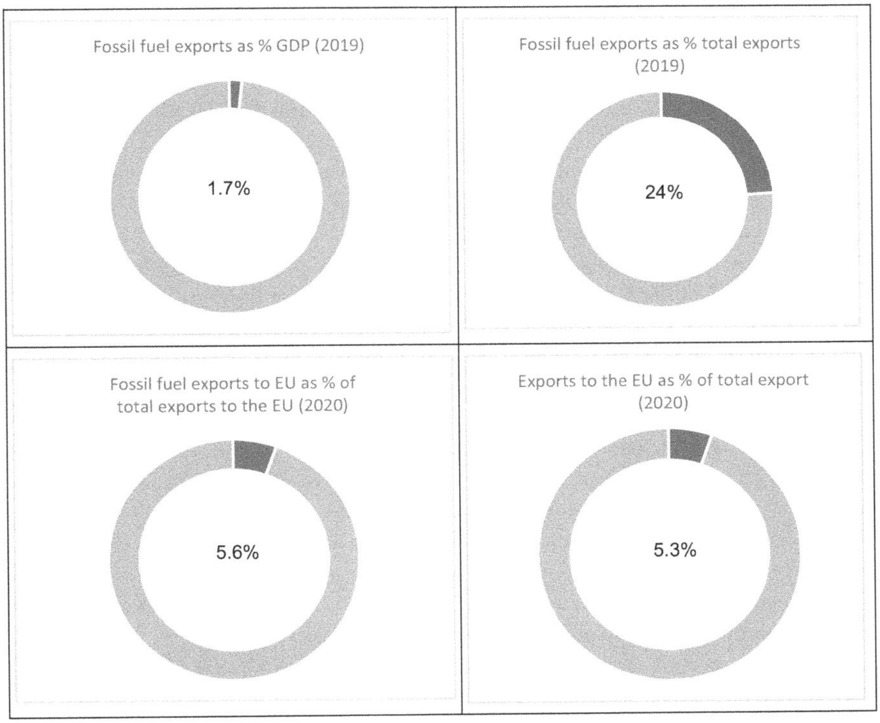

Figure 7.1 Status of fossil fuel exports of Canada in 2019/2020.

Sources: Data based on: https://data.worldbank.org (fossil fuel exports as a percentage of merchandise export; oil, gas, and coal as a percentage of GDP – accessed on 25 July 2021) and European Commission (2021a) (exports to EU – accessed on 25 July 2021).

the United States. Most of Canada's crude oil goes to the United States principally due to a lack of sufficient export capacity in Canada to send its liquids elsewhere. All of the country's current natural gas exports go to the United States. Canada also imported energy products from the United States worth CAN$35 billion in 2019 (Government of Canada 2020a; see also EIA 2019). The pressure to advance decarbonisation therefore depends to a significant degree also on the US demand for Canadian oil and gas.

Fossil fuels hence play an important role in Canada's economy and government budget. Especially oil and gas industries are an important engine of the economy, with the overall energy sector (including electricity) accounting for about 10 percent of GDP (fluctuating somewhat with the oil price). Fossil fuel exports accounted for less than 5 percent of GDP in 2019. The contribution of fossil fuels to the government budget was significant (but not dominant), accounting for more than 7 percent of all taxes paid by all industries over 2014–2018 averaging CAN$14 billion per year (Natural Resources Canada 2020). Government revenue for the fiscal year 2018–2019 was CAN$332.2 billion (Government of Canada 2019). The sector employs approximately 280,000 people directly (and about

550,000 indirectly), accounting for 1.5 percent of total employment (Natural Resources Canada 2020).

Different from other fossil fuel exporting countries, Canada possesses a highly developed and diversified economy. Fossil fuel production and export are significant (especially, as mentioned previously, in Alberta, Saskatchewan, and British Columbia), but the Canadian economy has other significant sectors to build on (see below).

Canada has also provided significant subsidies for fossil fuels. Data on these subsidies are hard to come by. Canada's federal Auditor General in 2017 expressed frustration at his inability to gain access to government documents which would allow him to determine the extent of the country's subsidies for the oil, gas, and coal industries (Office of the Auditor General of Canada 2017). The International Institute for Sustainable Development estimated that Canada's annual fossil fuel subsidies – including tax breaks and direct cash – in 2013–2015 amounted to CAN$3.3 billion annually. Tax expenditures occurring at both the federal and provincial levels represent a combined minimum total of CAN$2.5 billion annually (CAN$1.6 billion at federal level), while direct spending includes budgetary transfers the Canadian government provides to producers of oil, gas, and coal (Touchette 2015). The International Institute for Sustainable Development found in 2018 that federal subsidies had declined somewhat in 2016–2018 compared to 2015 but that this decline was not the result of a subsidy reform and may not be lasting (also because subsidies interrelate with oil prices) (IISD 2018a). The Canadian government has also drawn criticism because a significant amount of spending for the recovery from the Covid-19 crisis went to the production of fossil fuels (Climate Action Tracker 2021).

Canada is, as a result of the significance of its oil and gas sector, exposed to variations in oil and gas prices to some extent. Declining oil prices after 2014 have left a mark on economic development. GDP per capita in constant 2010 US$ stagnated from 2014 to 2016 (at approximately US$50,000) and rose slightly to about US$52,000 in 2019, before declining to US$48,600 in 2020 due to the Covid-19 crisis. GDP growth slowed from 2.9 percent in 2014 to 0.7 percent in 2015 and 1.0 percent in 2016. GDP growth picked up with the recovery of world oil prices in 2017 (3 percent) but fell again to 1.9 in 2019 before a recession of minus 5.4 percent in 2020 as a result of the Covid-19 pandemic (World Bank 2021c, 2021d).

Accordingly, the government budget situation has also seen significant fluctuations. Canada's public debt-to-GDP ratio rose from 85.7 percent in 2014 to 91.8 percent in 2016, before dropping slightly to 90.1 percent in 2017 and 86.8 percent in 2019 with the recovery of oil prices. The ratio jumped to 117.8 percent in 2020 as a result of the Covid-19 crisis (Trading Economics 2021). Given the diversification of the Canadian economy, oil price fluctuations have left their mark on economic development and the government budget but have overall remained manageable, as they could be balanced by other sectors of the economy.

Canada also harbours significant investments in oil and gas that may become 'stranded'. Reflecting its resource base, Canada has a vast network of more than 840,000 km of oil and gas pipelines. Pipelines mainly serve to transport oil, natural gas, and liquefied natural gas from Alberta west to British Columbia, north to the Northwest Territories, east to Quebec, and south to the United States (Texas). Several major additional pipeline projects are under construction or in planning, including the Trans Mountain expansion and

Enbridge's Line 3 replacement project. Over the years, indigenous groups, environmentalists, municipalities, mayors, and labour unions have increasingly opposed and legally challenged new pipeline projects over fears of contamination and climate change (Hughes 2018; The Canadian Encyclopedia 2018).

After years of political controversy, TransCanada's contentious Keystone XL project was abandoned by the developer in June 2021. In January 2021, during his first day in office, US President Joe Biden had revoked the permit for the pipeline issued by former president Donald Trump. The project would have significantly increased oil transport capacity to the United States at a cost of US$5–8 billion but faced severe criticism from environmentalists (Puko and Monga 2021).

Also beyond oil and gas pipelines, Canada is continuing to make considerable investments into fossil fuels. These in particular concern the production of oil from the large oil sands deposits, especially in Alberta. They also include investments in liquefied natural gas (LNG) terminals. In 2018, total Canadian energy assets amounted to CAN$685 billion (up 5 percent from 2017, mainly due to an increase in assets abroad in the United States and Mexico), of which assets worth CAN$452 billion (or 66 percent) were in Canada (Natural Resources Canada 2020).

The risk of stranded assets is far lower in the power sector. This is not least the result of the dominant role of hydropower in the Canadian electricity system. About two-thirds of Canada's total electricity production comes from renewables, and hydropower alone accounts for about 60 percent (with wind, biofuels, and solar providing for the balance). Hydropower generated 384,600 GWh in 2018, with an emphasis on Quebec, Manitoba, and Newfoundland and Labrador. Nuclear power provided a further 15 percent of total electricity in 2018. Coal and gas provided the remainder and were concentrated in selected provinces, contributing to electricity production in particular in Alberta, Saskatchewan, and Nova Scotia (Natural Resources Canada 2020). Canada is one of the co-founders of the Powering Past Coal Alliance and has committed to a phase-out of coal in electricity production by 2030. In 2018, it adopted performance standards for coal and natural gas-fired power stations to this end. Accordingly, it plans to significantly increase the use of renewables (including wind and solar, but also further expanding the use of hydropower and biomass) in power generation (Oil Change International 2015; Hughes 2018; Climate Action Tracker 2021).

Other fragility and security risks

As a highly developed country with a stable federal parliamentary democracy, Canada faces few other fragility risks. A range of available indices confirm that Canada is a stable federal parliamentary democracy with strong socio-political institutions, a welfare state with well-developed social security, public healthcare, and educational systems. Accordingly, Canada ranked 171 out of 179 countries on the 2021 Fragile States Index. Canada scores very highly on most indicators and has made significant progress on economic equality, while 'group grievances' remain somewhat of a challenge (especially reflecting internal challenges between francophone and anglophone communities) (Fund for Peace 2021). Corruption is not a significant issue with Canada ranked 11 among the 180 countries on Transparency International's Corruption Perceptions Index 2020 (Transparency

International 2021). Also, Canada ranks high on the Freedom House Index of 2021 scoring a near-perfect 98 aggregate score (highest mark in the political rights category) (Freedom House 2021) as well as on the World Bank's Worldwide Governance Indicators (World Bank 2021f; see also Figures 7.2 and 7.3).

As indicated, a particular feature of Canadian politics is longstanding differences and tensions between French-speaking and English-speaking parts. French-speaking Quebec hosts a strong independence movement. Two referendums on achieving (greater) independence from the rest of Canada were narrowly defeated in the 1980s and 1990s. Underlying tensions have since been successfully managed but remain a virulent undercurrent of Canadian politics.

Canada faces significant climate change impacts that are, however, comparatively modest. Climate change impacts are already being felt in Canada and set to grow further. Temperature increase in Canada exceeded 1.5°C between 1950 and 2010, approximately double the global average. This has had significant impacts in the coastal and northern

Fragility (2021)	SUSTAINABLE	Stable	Warning	Alert
Human development (2019)	VERY HIGH	High	Medium	Low
Strength of governance (2019)	VERY HIGH	High	Medium	Low
Climate change vulnerability (2019)	Low	**MEDIUM**	High	Very high
Sustainable energy development (2019)	VERY HIGH	High	Medium	Low

Figure 7.2 Governance indicators (Canada).
Sources: Fund for Peace (2021), ND-GAIN (2020), UNDP (2020), World Bank (2021f, 2021g).

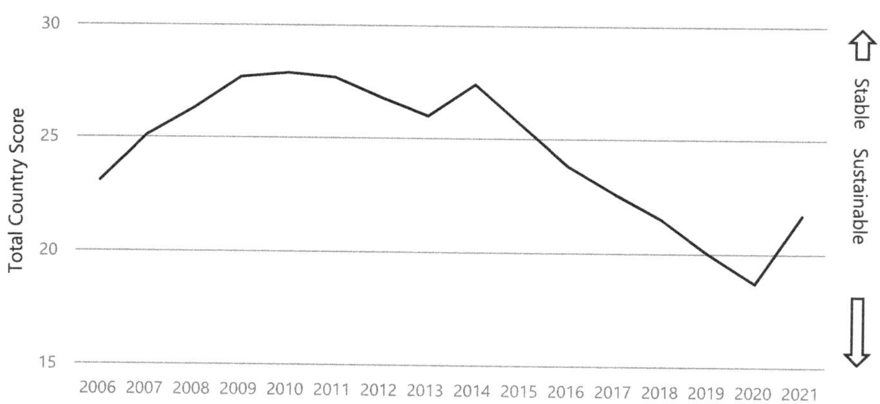

Figure 7.3 Canada's performance in the Fragile States Index 2006–2021.
Source: Fund for Peace (2021).

Note: Higher index numbers indicate a higher level of state fragility.

parts of the country. Indigenous peoples are the most vulnerable to climate change impacts. As they enjoy special rights in Canada, they have become increasingly important voices advocating action on climate change. Further impacts such as an increase in extreme weather events, sea-level rise, and a further change of the northern and Arctic parts of the country are expected. At the same time, some positive impacts such as a growth of food production from agriculture are also anticipated (Warren and Lemmen 2014; Government of Canada 2016b; Bush and Lemmen 2019).

Accordingly, Canada does not rank high on the Global Climate Risk Index for 2019 (60 out of 130) and for the period 2000–2019 (93 out of 130). The index assesses the extent to which countries have been subject to weather-related loss events (storms, floods, heat waves, etc.) (Germanwatch 2021). Similarly, the index of the Notre Dame Global Adaptation Initiative, which considers eight life-supporting sectors (food, water, health, ecosystems, human habitat, coastal, energy, and transportation infrastructure), gives Canada a favourable score and rank for its overall vulnerability and readiness. Its vulnerability is considered relatively low (the 12th least vulnerable country assessed in 2018) and its readiness high (among the top 20 in 2018) (ND-GAIN 2020; Figure 7.2). Hence, climate change impacts in Canada are considered significant but modest, and Canada possesses considerable means to manage and adapt to these impacts and hence is relatively resilient.

Past and present efforts to decarbonise

International climate policy

Canada is a party to the United Nations Framework Convention on Climate Change (UNFCCC) and its Paris Agreement. It ratified the UNFCCC in 1992 and the Paris Agreement in 2016. It was also originally a party to the Kyoto Protocol. It withdrew from the Protocol in late 2011, when it was clear that the country would not be able to live up to its commitment to reduce greenhouse gas (GHG) emissions by 6 percent below 1990 levels. Canada is an Annex I party to the UNFCCC and considered a developed country for the purposes of the climate treaties. Within the UN climate negotiations, Canada has formed part of the so-called Umbrella Group of non-EU developed countries.

Under the Paris Agreement, Canada submitted an original first Nationally Determined Contribution (NDC) in 2017 that committed it to achieving an economy-wide (including LULUCF – land use, land use change, and forestry) GHG emission reduction target of 30 percent below 2005 levels by 2030. This implied a reduction from 747 Mt CO_2 equivalent in 2005 to 523 Mt CO_2 equivalent in 2030 (Government of Canada 2017). In 2016, the Canadian government further adopted a long-term target of reducing emissions by 80 percent below 2005 levels by 2050. The 2017 submission in particular also provides information on the Pan-Canadian Framework on Clean Growth and Climate Change adopted in late 2016 as the main vehicle for the implementation of the Canadian NDC (see below on the status of domestic climate policy).

In July 2021, Canada submitted an update of its first NDC and increased its GHG emission reduction target (including LULUCF) to 40–45 percent below 2005 levels by 2030. Accordingly, Canadian GHG emission in 2030 would need to be reduced to 406.5–443.4 Mt CO_2 equivalent. At the same time, Canada also announced to aim at reaching climate neutrality (net zero emissions) by 2050. The 2021 submission provided supporting information on Canada's strengthened climate plan launched in December 2020 (Government of Canada 2021).

Both the initial Canadian NDC and its update represent a significant departure from a business-as-usual scenario that would have reduced projected Canadian GHG emissions to around 2005 levels by 2030. Measures have been taken since 2016, and further measures are required to achieve the envisaged 40–45 percent reduction. The Canadian government's own projections show that the national measures currently planned and under implementation will significantly reduce GHG emissions but will be insufficient to achieve the 2030 target. To this end, the Canadian government has identified a broad set of measures in its strengthened climate plan of 2020 (Government of Canada 2021).

The Climate Action Tracker concludes that the Canadian NDC has been and remains 'insufficient' for achieving the Paris Agreement's temperature goal of well below 2°C or 1.5°C of warming. However, the updated NDC brings Canada close to being '2°C compatible'. However, a significant implementation gap remains to be filled to actually achieve the upgraded NDC (Climate Action Tracker 2021).

GHG emissions

According to the UNFCCC, Canada's GHG emissions excluding LULUCF in 2019 stood at 730.245 Mt CO_2 equivalent (up 21.4 percent since 1990) (UNFCCC 2021). Given its large forestry sector, LULUCF is significant and was estimated to have acted as a carbon sink until 2018, however with shrinking negative emissions (from −41.63 Mt CO_2 equivalent in 2009 to −12.86 Mt CO_2 equivalent in 2018, according to UNFCCC data) (WRI 2021). In 2019, the sector was estimated to have emitted 9.878 Mt CO_2 equivalent (UNFCCC 2021). Accounting of LULUCF is surrounded by large uncertainties, as also illustrated by the fact that other sources have come to the conclusion that LULUCF has for long been a net source (WRI 2021) (see Figure 7.4).

Together, the oil and gas sectors were the largest GHG emitter in 2019 contributing about 26 percent to total Canadian emissions, followed by the transport sector with 25 percent. Other sectors like buildings, electricity, heavy industry, agriculture, and waste and others each accounted for between 7 and 12 percent of total GHG emissions (Environment and Climate Change Canada 2021). Between 1990 and 2019, emissions of the oil and gas and transport sectors rose by 87 percent and 54 percent, respectively. In contrast, emissions in the electricity and heavy industry sectors declined. In 2019, the GHG with the largest reported emissions (without LULUCF) was CO_2 (79.75 percent), while methane (CH_4), nitrous oxide (N_2O), and aggregate fluorinated GHG emissions stood at 13.39 percent, 5.01 percent, and 1.85 percent, respectively (Environment and Climate Change Canada 2021; UNFCCC 2021).

Climate policy

Under the then-new liberal Prime Minister Trudeau, Canada adopted a Pan-Canadian Framework on Clean Growth and Climate Change in December 2016 to (over)achieve Canada's original NDC to reduce GHG emissions to 30 percent below 2005 levels by

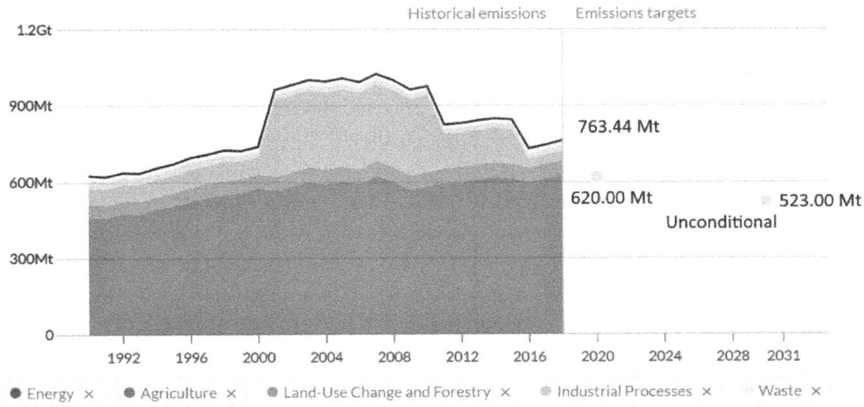

Figure 7.4 Canada's greenhouse gas emissions and emission targets (1990–2018).
Source: WRI (2021).

2030. This Framework constituted 'a comprehensive plan to reduce emissions across all sectors of the economy, accelerate clean economic growth, and build resilience to the impacts of climate change' (Government of Canada 2017). After its re-election, the liberal government released a strengthened climate plan for 'A Healthy Environment and a Healthy Economy', in December 2020 as a key plank to realise the ambition of the updated NDC to reduce GHG emissions by 40–45 percent below 2005 levels by 2030 (Government of Canada 2021).

The central pillar of modern Canadian climate policy has been the establishment of a carbon price across Canada, rising from an initial at least CAN$10 per tonne of CO_2 equivalent to CAN$50 per tonne by 2022. To be implemented in collaboration with the Canadian provinces that enjoy considerable flexibility as to how to implement carbon pricing (e.g., through a tax or levy or through emissions trading), the carbon pricing initiative saw a delay to 2019 and opposition in particular from the conservative opposition party. Victories for the conservatives in the provincial elections in Ontario in 2018 and Alberta in 2019 have led to rollbacks of climate policy in these provinces. Court proceedings against carbon pricing led to its reconfirmation in 2021, when the supreme court declared the initiative constitutional. The strengthened climate plan of December 2020 foresees an increase of the carbon price by CAN$15 per year from 2023 reaching CAN$170 by 2030 (Government of Canada 2017, 2021; Climate Action Tracker 2021).

The Pan-Canadian Framework and the strengthened climate plan contain a number of complementary measures beyond the carbon pricing initiative. Hence, Canada is to develop a clean fuel standard to reduce emissions from fuels used in transportation, buildings, and industry and drive down emissions from electricity, including through phasing out traditional coal units by 2030 and performance standards for natural gas-fired electricity, and modernising Canada's electricity systems. New codes for new and existing buildings and new standards for energy-efficient appliances and equipment are planned. In the transport sector, planned actions include the aim of all cars and vans sold to be zero emission by 2035 and promoting the shift to lower-emitting transportation and using cleaner fuels. The framework also aims to achieve a 40–45 percent reduction of methane emissions from the oil and gas sectors by 2025 through regulations and covers actions to enhance carbon sinks (forests, wetlands, and agricultural lands), waste-to-energy, and adaptation. Throughout the different sectors, targeted investments and incentives play a particularly prominent role (Government of Canada 2017, 2021; see also Climate Action Tracker 2021).

Climate investments have been further boosted and reinforced by the response to the Covid-19 crisis. To be sure, the overall record of the Canadian recovery spending has been mixed because significant support also went to fossil fuel industries. However, a comparatively large part of recovery spending has been found to have been allocated for clean energy and, more generally, for greening (Climate Action Tracker 2021; Energy Policy Tracker 2021; Global Recovery Observatory 2021).

Natural and man-made sinks play an important role in Canada's climate plans. Canada has already made significant efforts in harnessing Carbon Capture and Storage (CCS) and continues to do so. It harbours several large-scale operational CCS projects and continues to support and foster CCS. In addition, enhancing natural carbon sinks, prominently including forests, constitutes an important focus of its climate plans (Natural Resources Canada 2013; Government of Canada 2021).

In line with the overall federal governance system, Canadian provinces and territories enjoy considerable flexibility to determine policies, including climate policy (see above). As a result, both Quebec and Ontario established emissions trading systems, which were linked to each other and to the emissions trading system of California through the Western Climate Initiative. However, a change of government in Ontario in 2018 resulted in the province abandoning its emissions trading system in 2018 (Climate Action Tracker 2021).

Climate politics

The future of Canadian climate policy is also uncertain, because climate change and climate policy has remained a very partisan issue in Canadian politics. As a result, Canada has not been able to maintain a stable trajectory in its climate policy over the years, with changes of government between liberal and conservative resulting in major shifts in climate policy. Whereas liberal governments have tried to advance Canadian climate policy, conservative governments have stalled progress and have at times actively undermined climate action. The liberal government of the 1990s signed and ratified the Kyoto Protocol, which saw a commitment to reducing emissions by 6 percent from 1990 levels between 2008 and 2012 (but failed to introduce a GHG reduction plan). The subsequent government of Conservative Prime Minister Stephen Harper, in power from 2006 to 2015, abandoned Canada's commitment and instead advanced policies that made emissions rise to a new all-time high of 748 Mt CO_2 equivalent (CBC News 2007). In 2011, Canada became the first party to withdraw from the Kyoto Protocol. The pendulum swung back with the change of government to liberal Prime Minister Justin Trudeau in 2015. Action on climate change had been a prominent part of his election and re-election platforms, which were followed up most notably by establishing and implementing the aforementioned Pan-Canadian Framework and strengthened climate plan of 2020. Nonetheless,

Trudeau has also maintained firm support for oil pipelines and the creation of a liquefied fracked gas export industry, despite opposition from environmental groups and sections of the public. The liberal Canadian government furthermore established a Just Transition Task Force in 2018 to consider and mitigate negative impacts of the energy transition on workers and affected communities with participation of stakeholders (IISD 2018b; see also Government of Canada 2021: 9–10).

As mentioned above, provinces and territories enjoy significant room for political manoeuvre in the Canadian federal system, including on climate policy. Some provinces have advanced more ambitious climate policies than the federal government. For instance, Alberta is planning to phase out coal electricity by 2030 and cap overall oil sands emissions, while Quebec and British Columbia have already enacted policy instruments including phasing out coal, implementing a carbon tax, and working with California on a regional cap-and-trade system (Government of Canada 2016a, 2021). As at the federal level, climate policy is frequently advanced under liberal provincial governments, while at times being turned back under conservative leadership (see on Ontario above).

Public opinion in Canada has displayed awareness of climate change but has remained somewhat divided when it comes to acknowledging that this change is caused by humans and with respect to climate policy. Overall, it has favoured climate action despite the significant energy resources that the country owns. According to a public survey conducted by Yale University's Programme on Climate Change Communication (dataset of more than 5,000 individuals and collected since 2011), 79 percent of Canadians believe that the Earth is getting warmer and 61 percent believed that the Earth is getting warmer partly or mostly (44 percent) due to human activities. As much as 66 percent of Canadians interviewed were in favour of a cap-and-trade system and 49 percent believed that taxes on carbon-based fuels should be increased. Significant differences exist between different provinces, with belief in climate change clearly less in Alberta, Saskatchewan, and Manitoba (Mildenberger et al. 2016).

Overall, Canadian climate policy has effectuated a significant deviation from business as usual, but its ability to deliver on the 2030 NDC remains uncertain at best. Important measures have been initiated under a Pan-Canadian Framework on Clean Growth and Climate Change and the strengthened climate plan of 2020. However, they fall short of what would be required and the effects on Canadian GHG emissions have so far been limited. The achievement of Canada's NDC is also uncertain as climate policy has remained a partisan issue in Canadian politics, with the conservative party largely opposed to decisive climate action. Hence, it is as yet uncertain whether Canada will be able to fully achieve its 2030 target.

Trends and potential

Canada possesses most of the means required to wean itself off fossil fuels. It has a developed educational system and a strong human capital resource base. It also possesses a developed economy with an evolving services sector and a strong innovation potential. And its natural resource endowments include plentiful renewable energy resources that should enable the decarbonisation of the Canadian economy and provide strong opportunities for exporting renewable energy, especially to north-eastern US states.

Education and knowledge-based economy

Canada ranks high on education and skills. According to the World Bank's Human Capital Index, the current level of education and healthcare would allow the average child born in Canada today to achieve about 80 percent of its potential, clearly above the worldwide average and ranking under the top ten of 174 countries (World Bank 2020). Canada spends more on education as a percentage of GDP than the OECD average and also has a higher-than-average share among OECD countries of 25–34 year olds who have a tertiary qualification (63 percent as compared with 45 percent) and has one of the lowest unemployment rates in the OECD. At the same time, the percentage of first-time master's graduates younger than 35 is relatively low. There is therefore also room for further improvement (OECD 2018, 2020).

Growing services sector

The Canadian services sector is well developed, accounting for 70 percent of the Canadian GDP and employing three quarters of the Canadian working population. Among the services sector, the retail sector employs the largest percentage of Canadians (12 percent) followed by business services (which include financial services, real estate, and communications industries), and the health and education sectors (both largely under the influence of the government). Canada also has a significant high-tech industry, and a burgeoning film, television, and entertainment industry, in addition to a growing tourism industry. Canada also ranks high in terms of innovation. The 2018 Global Innovation Index ranked Canada 18th globally and the top innovation region in the world along with the United States (Cornell University, INSEAD, and WIPO 2018).

In the 2010s, the services sector overtook the manufacturing and energy sector as the engine of development and job creation (Kirby 2017). Since 2013, the vast majority of jobs have been created in the services sector, while there have been fewer jobs in the oil and gas sector. The case is the same in energy-rich provinces like Ontario and Alberta. In 2016, the goods sector in Ontario generated 5,000 jobs as compared to 75,000 in the services sector, while in Alberta, the services sector added 34,000 jobs as opposed to a loss of 53,000 jobs in the manufacturing sector (Kirby 2017). Overall, the Canadian economy therefore has well-developed potential to transition away from the production of fossil fuels and towards a low-carbon economy. In this context, it should be highlighted that the 'old' fossil fuel-based industries are concentrated in a few provinces and regions, requiring particular attention when devising accompanying measures to smooth the transition. The long-term impacts of the Covid-19 pandemic are as yet unclear.

Renewable energy and clean energy

Canada also possesses considerable potential to further develop the use of renewable energy, especially in the power sector. As mentioned before, about two-thirds of Canada's power generation already comes from renewable sources, particularly hydropower that alone accounts for around 60 percent. The country has very significant potential to further develop renewable power generation, including in the provinces that currently rely more heavily on fossil fuels (such as Alberta and Saskatchewan). According to Canada's Mid-Century Long-Term Low-Greenhouse Gas Development Strategy, electricity generation should rise from 17 percent of delivered energy in 2016 to 33–65 percent in 2050 (Hughes 2018). This increase could be supplied through increases in renewable energy, including further hydropower and a strong increase in the use of wind and solar energy as well as biomass. Accordingly, the National Energy Board of Canada projected in 2017 that Canada's renewable capacity could grow rapidly by 2040, with wind power doubling or even tripling and solar power potentially increasing more than tenfold (National Energy Board 2017). The 2050 strategy foresees even higher potential for increases, including in biomass and hydropower, which will require very significant investments (Hughes 2018). In this context, the very large availability of hydropower resources in Canada seems ideal for the decarbonisation of the power system in a mix with more intermittent renewable energy sources such as wind and solar. The 2020 strengthened climate plan and recovery spending already aim at further developing Canada's clean energy potential,

The considerable renewables potential in Canada also creates opportunities for related exports, especially to the north-eastern states of the United States. Greater exports of zero-carbon electricity to the north-eastern Unites States would require the construction of long-distance transmission lines to US markets. As a result, Canada could also provide its hydroelectric reservoirs as enormous energy storage services. Such an arrangement could reduce the cost of decarbonisation in the US northeast making both the United States and Canada more energy-competitive (Sachs 2018). This could help balance future declines of fossil fuel exports to the United States.

Cooperation with the EU

EU-Canada relations have been close and dense for several decades, as evidenced by the conclusion of the Comprehensive Economic and Trade Agreement (CETA) in 2016, as well as long-established close cooperation in a multitude of settings and areas. Trade and investment as well as energy (both fossil fuels and renewables) are important and evolving areas, with significant potential to integrate decarbonisation. Historical ties and shared interests and values provide a fruitful basis for developing EU-Canada relations under decarbonisation and help both partners address shared challenges in this respect. Overall, there is a firm basis to further advance EU-Canada relations independently from fossil fuels grounded in shared values.

Institutional framework: Comprehensive Economic and Trade Agreement and Strategic Partnership Agreement

The EU (then the European Economic Community) concluded a first Framework Commercial and Economic Agreement with Canada in 1976, also opening its overseas diplomatic mission in Ottawa. The Framework Agreement was the first of its kind with an industrialised country. The EU-Canada relationship is based on shared values such as democracy, fundamental human rights, rule of law, free trade, and multilateralism. Cooperation between these likeminded partners has evolved into a firm, strategic partnership with a balanced focus on both socio-economic and politico-security cooperation. Annual high-level bilateral summits bring together the Canadian Prime Minister, the European Council and Commission Presidents, and the High Representative for Foreign Affairs and Security Policy allowing a regular stocktake and reiteration of bilateral ties. Since the early 2000s, Canada has also been acknowledged as one of the EU's ten strategic partner countries (along with the United States, Mexico, Brazil, Russia, India, South Africa, China, Japan, and South Korea).

In this context, the partners in 2004 adopted an ambitious strategic partnership agenda and agreed on a framework for a new Canada-EU Trade and Investment Enhancement Agreement. The latter evolved into CETA. CETA negotiations officially began in 2009 and were concluded in August 2014. CETA was signed in 2016 and provisionally applied as of September 2017. It would eliminate 98 percent of tariffs between both parties. Alongside CETA, the EU and Canada launched a Strategic Partnership Agreement which sought to deepen and broaden political cooperation on a series of issues like international peace and security, counterterrorism, human rights and nuclear non-proliferation, clean energy and climate change, migration and peaceful pluralism, sustainable development, and innovation. The Partnership Agreement was signed alongside CETA in 2016 and subsequently provisionally applied from 1 April 2017. Both partners have been active international proponents of free trade and anti-protectionism.

Trade and investment

Trade relations between the EU-27 and Canada are strong. In 2019, bilateral trade in goods between both partners stood at €59.3 billion (excluding the United Kingdom). Canada was the EU's 11th top trading partner representing about 1.5 percent of total EU external trade, while the EU was Canada's third largest trading partner after the United States and China, accounting for more than 8 percent of its global trade. EU exports to Canada amounted to €38.3 billion (accounting for about 11 percent of Canadian imports – ranked third after the United States and China), while imports from Canada totalled €20 billion (accounting for more than 5 percent of Canadian exports – ranked second after the United States). This resulted in an EU trade surplus of about €18 billion. Trade in services exceeded €35 billion, also with a significant surplus for the EU. The stock of foreign direct investments between the two stood at about €640 billion, of which the EU accounted for nearly €400 billion. In the aftermath of CETA, trade volumes have grown (while Brexit reduced trade volumes by more than 20 percent). The year 2020 saw, however, a contraction especially of EU exports to Canada in the wake of the Covid-19 pandemic (European Commission 2021a, 2021b; see also EPRS 2017).

Trade in fossil fuels is relatively modest for both sides. In 2019, fuels accounted for about 7 percent of all EU exports to Canada, while EU fuel imports from Canada were about 10 percent of the total. As a result, trade in fuels between both partners is – with some annual variation – in balance. Over recent years, the top three categories of goods traded between the EU and Canada have been machinery (around 25 percent of both EU exports and imports), chemical and pharmaceutical products (more than 15 and close to 10 percent, respectively), and transport equipment (more than 15 percent and about 7 percent, respectively). Hence, fossil fuels represent a significant share of bilateral trade, but they are not the most important commodity (European Commission 2021a, 2021b; see also Figure 7.1).

Clean energy and climate change

Beyond existing relations related to trade and investment (including oil and gas – see above), the potential for cooperation on (renewable and clean) energy remains to be exploited to a larger extent. The EU and Canada established a High-Level Dialogue on Energy in 2007 and further refreshed it at the bilateral summit in 2014 (European Commission 2018). The dialogue covers areas of energy security, energy efficiency, renewable energy, safe and responsible exploitation of conventional and unconventional energy resources, LNG trade, and improved market transparency and stability. Specific issues discussed are split almost equally between renewable energy, fossil fuels, and international cooperation (international fora, Ukraine, global energy markets).

Both partners discuss renewable energy development domestically as well as the public acceptance of new energy infrastructure. Decarbonisation technologies like CCS and smart grids are also discussed through the perspective of research and innovation. A significant part of bilateral discussions continues to focus on gas and oil markets in Canada and the EU (including infrastructure, LNG, unconventional resources) and the evolving global LNG market. Bilateral cooperation on renewable energy remains limited to dialogues and workshops and has potential for further substantiation. However, CETA can help facilitate and boost trade in clean technologies given that it eliminates all tariffs on clean technology products traded between the partners and extends access to public procurement markets that can play a significant role in supporting the EU's and Canada's initiatives towards decarbonisation.

General geopolitics and broader cooperation agenda

Overall and beyond the areas addressed above, the EU and Canada share a very broad agenda and related interests and values that have grown historically. At a very general level, developing relations with the EU is one important way for Canada to balance its dependence on the United States and allows its voice to resonate internationally. Conversely, Canada is also an important partner for the EU and its member states in forming international alliances on a par with the United States and China.

The EU and Canada therefore cooperate in a variety of international institutions and on a number of foreign policy issues. Canada and the EU share similar positions on a number of critical international issues, including security (with close cooperation long established in the North Atlantic Treaty Organization [NATO]), human rights, free trade, and others. Both partners collaborate closely within the ambit of organisations like the UN, G7, G20, WTO, OECD, IAEA, and others.

Climate diplomacy has become an increasingly important facet of EU-Canada relations, especially following the decision by then US President Donald Trump to withdraw from the Paris Agreement in 2017. In September 2017, the EU, Canada, and China established a Ministerial on Climate Action to fill the void left by the United States under President Trump abandoning its convening of the Major Economies Forum on Energy and Climate.

The Ministerial on Climate Action has been convened annually since. Climate change was also a priority in the unsuccessful Canadian campaign for a seat on the UN Security Council in 2021–2022. Furthermore, Canada is a member of the Powering Past Coal Alliance, along with many EU member states.

Science and technology cooperation between the EU and Canada is also well developed and institutionalised. In 1996, both partners signed the bilateral Agreement for Scientific and Technological Cooperation, which is overseen by a Joint Science and Technology Cooperation Committee that meets regularly. The Agreement covers areas such as aeronautics, the bioeconomy, marine and arctic research, research infrastructures, health, nuclear and other areas. In 1998, Canada and Euratom signed an agreement on cooperation in nuclear research. Canada has also been an active participant in various of the EU research and innovation funding programmes, including the EU framework programmes for research and Marie Sklodowska-Curie Actions.

Conclusions

Fossil fuels (oil and gas, and, to a lesser extent, coal) are important sectors of the Canadian economy and provide for a significant share of Canadian exports. Hence, decarbonisation in Europe and the rest of the world requires Canada to adapt significantly, depending especially on developments in the United States as the major export market. This constitutes a major challenge for the Canadian economy and political system. Addressing this challenge requires consistent long-term effort to phase out fossil fuel production (and consumption), while also nurturing the country's many non-fossil-fuel-dependent economic sectors and activities. This challenge is reinforced by the fossil fuel economy being concentrated in just a few provinces (notably, Alberta, Saskatchewan, and British Columbia) – provinces that enjoy high levels of authority and independence from the federal government in Canada's federal political system. Achieving significant emission cuts will therefore depend on astute political decision-making and management to ensure that climate policy also takes into account the just transition to a low-carbon economy in more fossil-fuel-dependent regions of Canada.

Fragility risks present are relatively moderate in Canada. Canada is a highly developed country with a stable federal parliamentary democracy. Adverse impacts of climate change are comparatively modest, and Canada possesses sufficient resources to adapt to the impacts of the warming climate. However, northern parts of the country and indigenous peoples are most vulnerable to climate change impacts. With indigenous peoples enjoying particular rights in Canada, they have become increasingly prominent voices in favour of decisive climate action.

EU-Canada relations are not heavily based on fossil fuels and are already broadly anchored and very close. Fossil fuels account for a modest share of the significant trade relationship between the two partners. The EU and Canada already cooperate closely in a wide range of areas, including security and defence (within NATO), science and technology, clean and renewable energy, and climate diplomacy. As partners at a comparable level of development, with close historical ties, and similar political systems and values, both face similar challenges, also with respect to decarbonisation, and have a deeply rooted joint interest in developing the relationship. The EU is an important partner for Canada to balance its dependence on the United States, and Canada is important for the EU in its attempts to build international alliances. Cooperating allows both partners to boost their international standing. As such, there is ample scope for EU-Canada relations to flourish as they and the rest of the world decarbonise and for these relations to actively support the transition to a low-carbon economy in both jurisdictions.

Under the current circumstances, EU foreign policy towards Canada does not need to profoundly reorient under decarbonisation. Given the richness of contemporary EU-Canada relations and the limited role of fossil fuels in the relationship, the global shift towards decarbonisation should focus the EU and Canada's relations on developing areas of cooperation that will help both partners to reduce emissions in their respective jurisdictions. Key priorities in this regard include:

- Both the EU and Canada face the challenge of ensuring **a 'just transition' for coal/fossil fuel-producing regions**. In Canada, this challenge relates to particular provinces and regions, while in the EU it is particular member states and regions that are affected. Consequently, on both sides of the Atlantic a variety of approaches are currently being tried and tested to manage the transition away from fossil fuel production and generation. Thus, there is obvious potential for both partners to enter into a dialogue to identify best practices and further develop them in a mutual learning process. Related efforts could for example be embedded in broader existing frameworks, such as the Powering Past Coal Alliance.
- Beyond a focus on fossil-fuel-producing regions, Canada and the EU also face common challenges with regard to **other sectors and aspects of the climate and energy transition**. These include decarbonising the building and transport sectors, as well as energy-intensive industries, and expanding the use of renewable energy sources (including the related challenges of grid management and development). These common challenges provide fertile ground for targeted dialogues and for identifying specific opportunities and projects for cooperation, including joint research and development.
- In addition, Canada and the EU have considerable scope to further coordinate their **international policies** to support and raise ambition for decarbonisation. Both partners share important international objectives, including supporting the implementation and development of the Paris Agreement and a related international trade agenda. They can build on and further expand existing cooperation in this field (for example, in convening the Ministerial on Climate Action) to jointly build broader international alliances. In the context of Covid-19 pandemic, this includes efforts to coordinate towards a green recovery.

Overall, there is already ample scope for EU-Canada relations to develop fruitfully under decarbonisation. A joint climate agenda has the potential to advance the relationship in a targeted way, helping both partners to more effectively address the common challenges they face in the climate and energy transition, both domestically and at the international level. Decarbonisation therefore does not call for the EU and Canada to drastically recalibrate their relations; rather, it provides a strong rationale for focusing their attention on the common challenges they face in transitioning to a low-carbon economy and on deepening cooperation accordingly. Given the partisan nature of climate policy in Canada, developments in domestic Canadian politics will be a major factor influencing the prospects for deepening this cooperation in future. The outcome of the federal elections in October 2019 have therefore reinforced the basis for further advancing EU-Canada cooperation in this area in the early 2020s.

Note

1 Production, consumption, and export data vary between different sources (including IEA, CIA, BP Statistical Review) but all give roughly the same overall picture.

References

BP (2020) *BP Statistical Review of World Energy*. Available at: https://www.bp.com/content/dam/bp/business-sites/en/global/corporate/pdfs/energy-economics/statistical-review/bp-stats-review-2020-full-report.pdf (accessed 18 June 2021).

Bush, E. and Lemmen, D.S. (eds.) (2019) *Canada's Changing Climate Report*. Ottawa: Government of Canada. Available at https://changingclimate.ca/site/assets/uploads/sites/2/2020/06/CCCR_FULLREPORT-EN-FINAL.pdf.

CBC News (2007) 'Harper's letter dismisses Kyoto as "socialist scheme"', *CBC News*, 30 January 2007. Retrieved 22 April 2020 from https://www.cbc.ca/news/canada/harper-s-letter-dismisses-kyoto-as-socialist-scheme-1.693166.

Chatham House (2021) 'Resources Trade. Earth'. Retrieved 7 July 2021 from https://resourcetrade.earth/.

CIA – Central Intelligence Agency (2021) 'The World Factbook. Canada'. Retrieved 7 July 2021 from https://www.cia.gov/the-world-factbook/.

Climate Action Tracker (2021) 'Canada'. Retrieved 15 July 2021 from https://climateactiontracker.org/countries/canada/.

Cornell University, INSEAD and WIPO (2018) *The Global Innovation Index 2018: Energizing the World with Innovation*. Ithaca, NY, Fontainebleau and Geneva. Available at: https://www.wipo.int/edocs/pubdocs/en/wipo_pub_gii_2018.pdf (accessed 22 April 2020).

EIA – Energy Information Agency (2019) 'Canada'. Retrieved 25 June 2021 from https://www.eia.gov/beta/international/analysis.php?iso=CAN.

Energy Policy Tracker (2021) 'Canada'. Retrieved 15 July 2021 from https://www.energypolicytracker.org/country/canada.

Environment and Climate Change Canada (2021) *Canadian Environmental Sustainability Indicators: Greenhouse Gas Emissions*, April 2021. Available at: https://www.canada.ca/content/dam/eccc/documents/pdf/cesindicators/ghg-emissions/2021/greenhouse-gas-emissions-en.pdf (accessed 15 July 2021).

EPRS – European Parliamentary Research Service (2017) *Canada: Economic Indicators and Trade with EU*. Available at: https://www.europarl.europa.eu/RegData/etudes/ATAG/2017/595898/EPRS_ATA(2017)595898_EN.pdf (accessed 22 April 2020).

European Commission (2018) 'Canada'. Available at: https://ec.europa.eu/energy/topics/international-cooperation/key-partner-countries-and-regions/canada_pl (accessed 19 October 2020).

European Commission (2021a) 'European Union, Trade in Goods with Canada'. Retrieved 14 July 2021 from https://webgate.ec.europa.eu/isdb_results/factsheets/country/details_canada_en.pdf.

European Commission (2021b) 'Countries and Regions. Canada'. Retrieved 1 July 2021 from https://ec.europa.eu/trade/policy/countries-and-regions/countries/canada/.

Freedom House (2021) 'Freedom in the World 2021: Canada'. Retrieved 17 June 2021 from https://freedomhouse.org/country/canada/freedom-world/2021.

Fund for Peace (2021) 'Fragile States Index 2021'. Retrieved 26 July 2021 from https://fragilesstatesindex.org/.

Germanwatch (2021) 'Global Climate Risk Index 2021'. Retrieved 17 June 2021 from https://germanwatch.org/en/19777.

Global Recovery Observatory (2021) 'Canada'. Retrieved 15 July 2021 from https://recovery.smithschool.ox.ac.uk/tracking/.

Government of Canada (2016a) *PanCanadian Framework on Clean Growth and Climate Change: Canada's Plan to Address Climate Change and Grow the Economy*. Available at: https://www.canada.ca/content/dam/themes/environment/documents/weather1/20161209-1-en.pdf (accessed 22 April 2020).

Government of Canada (2016b) 'Pan-Canadian Framework on Clean Growth and Climate Change: Complementary Actions to Reduce Emissions'. Retrieved 22 April 2020 from https://www.canada.ca/en/services/environment/weather/climatechange/pan-canadian-framework/complementary-actions-reduce-emissions.html.

Government of Canada (2017) *Canada's Nationally Determined Contribution Submission to the United Nations Framework Convention on Climate Change*. Available at: https://www4.unfccc.int/sites/ndcstaging/PublishedDocuments/Canada%20First/Canada%20First%20NDC-Revised%20submission%202017-05-11.pdf (accessed 7 July 2021).

Government of Canada (2019) *Annual Financial Report of the Government of Canada Fiscal Year 2018–2019*. Available at: https://www.canada.ca/en/department-finance/services/publications/annual-financial-report/2019.html (accessed 29 June 2021).

Government of Canada (2020a) 'Energy and the Economy'. Retrieved 18 June 2021 from https://www.nrcan.gc.ca/science-data/data-analysis/energy-data-analysis/energy-facts/energy-and-economy/20062.

Government of Canada (2020b) 'How Does the Forest Industry Contribute to Canada's Economy?' Retrieved 1 July 2021 from https://www.nrcan.gc.ca/our-natural-resources/forests-forestry/state-canadas-forests-report/forest-industry-contribute/16517.

Government of Canada (2021) *Canada's 2021 Nationally Determined Contribution under the Paris Agreement*. Available at: https://www4.unfccc.int/sites/ndcstaging/PublishedDocuments/Canada%20First/Canada's%20Enhanced%20NDC%20Submission_EN.pdf (accessed 15 July 2021).

Hughes, D.J. (2018) *Canada's Energy Outlook: Current Realities and Implications for a Carbon-Constrained Future*. Prepared on behalf of the Canadian Centre for Policy Alternatives. Available at: https://ccpabc2018.files.wordpress.com/2018/05/cmp_canadas-energy-outlook-2018_full.pdf (accessed 22 April 2020).

IEA – International Energy Agency (2021) 'Canada. Data Browser. Electricity and Heat. Electricity Generation by Source'. Retrieved 28 June 2021 from https://www.iea.org/countries/canada.

IISD – International Institute for Sustainable Development (2018a) *Public Cash for Oil and Gas: Mapping Federal Fiscal Support for Fossil Fuels*. Available at: https://www.iisd.org/sites/default/files/publications/public-cash-oil-gas-en.pdf (accessed 22 April 2020).

IISD (2018b) *Real People, Real Change: Strategies for Just Energy Transitions*. Available at: https://www.iisd.org/system/files/publications/real-people-change-strategies-just-energy-transitions.pdf (accessed 22 April 2020).

Kirby, J. (2017) 'Goods and services: Canada's other two-speed economy', *Maclean's*, 6 January 2017. Retrieved 22 April 2020 from https://www.macleans.ca/economy/economicanalysis/canadas-other-two-speed-economy/.

Mildenberger, M., Howe, P., Lachapelle, E., Stokes, L., Marlon, J., and Gravelle, T. (2016) 'The distribution of climate change public opinion in Canada'. Available at SSRN. https://dx.doi.org/10.2139/ssrn.2732935.

National Energy Board of Canada (2017) *Canada's Energy Future 2017: Energy Supply and Demand Projections to 2040*. Available at: https://www.cer-rec.gc.ca/en/data-analysis/canada-energy-future/2017/2017nrgftr-eng.pdf (accessed 22 April 2020).

Natural Resources Canada (2013) *Carbon Capture and Storage: Canada's Technology Demonstration Leadership*. Available at: https://www.nrcan.gc.ca/sites/www.nrcan.gc.ca/files/energy/files/pdf/11-1416_eng_acc.pdf (accessed 1 July 2021).

Natural Resources Canada (2019) 'Energy Maps'. Retrieved 18 June 2021 from https://www.nrcan.gc.ca/maps-tools-publications/tools/geodetic-reference-systems/energy-maps/16872#non-renewableresources.

Natural Resources Canada (2020) *Energy Fact Book – 2020–2021*. Available at: https://www.nrcan.gc.ca/sites/nrcan/files/energy/energy_fact/energy-factbook-2020-2021-English.pdf (accessed 18 June 2021).

ND-GAIN – Notre Dame Global Adaptation Initiative (2020) 'ND-GAIN Country Index'. South Bend, IN: University of Notre Dame. Retrieved 5 July 2021 from https://gain.nd.edu/our-work/country-index/.

OECD – Organization for Economic Co-operation and Development (2018) 'OECD GPS: Canada Overview of the Education System (EAG 2017)'. Retrieved 22 April 2020 from https://gpseducation.oecd.org/CountryProfile?primaryCountry=CAN&treshold=10&topic=EO.

OECD (2020) *Education at a glance 2020: OECD indicators*. Paris: OECD Publishing. Available at: https://doi.org/10.1787/69096873-en (accessed: on 15 July 2021).

Office of the Auditor General of Canada (2017) 'Spring Reports of the Auditor General of Canada to the Parliament of Canada, Report 7 – Fossil Fuel Subsidies'. Retrieved 22 April 2020 from https://www.oag-bvg.gc.ca/internet/English/parl_oag_201705_07_e_42229.html#.

Oil Change International (2015) *On the Edge: 1.6 Million Barrels per Day of Proposed Tar Sands Oil on Life Support*. Available at: http://priceofoil.org/content/uploads/2015/05/OCI-Briefing-OnTheEdge_FINAL+.pdf (accessed 22 April 2020).

Puko, T. and Monga, V. (2021) 'Keystone XL oil project abandoned by developer. Decision ends a year's long effort to pipe more Canadian crude to the U.S.' *The Wall Street Journal*, 9 June 2021. Retrieved 30 June 2021 from https://www.wsj.com/articles/keystone-xl-oil-project-abandoned-by-developer-11623272010.

Sachs, J. (2018) 'Forget Trans Mountain, here's the sustainable way forward for Canada's energy sector', *The Globe and Mail*, 13 April 2008. Retrieved 22 April 2020 from https://www.theglobeandmail.com/opinion/article-the-sustainable-way-forward-for-canadas-energy-sector/.

Statistics Canada (2016) '2016 Census: 150 Years of Urbanization in Canada'. Retrieved 22 April 2020 from https://www.statcan.gc.ca/eng/sc/video/2016census_150yearsurbanization.

The Canadian Encyclopedia (2018) 'Pipelines in Canada'. Retrieved 22 April 2020 from https://www.thecanadianencyclopedia.ca/en/article/pipeline.

Touchette, Y. (2015) *G20 Subsidies to Oil, Gas and Coal Production: Canada*. Overseas Development Institute. Available at: https://cdn.odi.org/media/documents/9988.pdf (accessed 22 April 2020).

Trading Economics (2021) 'Canada Government Gross Debt to GDP'. Retrieved 17 June 2021 from https://tradingeconomics.com/canada/government-debt-to-gdp.

Transparency International (2021) 'Corruption Perceptions Index 2020'. Retrieved 17 June 2021 from https://www.transparency.org/en/cpi/2020/index/nzl.

UN DESA – United Nations Department of Economic and Social Affairs (2019) 'World Population Prospects (2019): Total Population by Sex (Thousands). Canada'. Retrieved 17 June 2021 from https://population.un.org/wpp/DataQuery/.

UNDP – United Nations Development Programme (2020) 'Human Development Report'. Available at: http://hdr.undp.org/en/countries (accessed 26 July 2021).

UNFCCC – United Nations Framework Convention on Climate Change (2021) 'Summary of GHG Emissions for Canada'. Retrieved 1 July 2021 from https://di.unfccc.int/ghg_profiles/annex-One/CAN/CAN_ghg_profile.pdf.

World Bank (2020) '2020 Human Capital Index: Country Briefs and Data'. Retrieved 1 July 2021 from https://www.worldbank.org/en/publication/human-capital#Index.

World Bank (2021a) 'Exports of Goods and Services (Current US$) – Canada'. Retrieved 18 June 2021 from https://data.worldbank.org/indicator/NE.EXP.GNFS.CD?locations=CA.

World Bank (2021b) 'GDP (Current US$)'. Retrieved 17 June 2021 from https://data.worldbank.org/indicator/NY.GDP.MKTP.CD?locations=CA.

World Bank (2021c) 'GDP Growth (Annual%) – Canada'. Retrieved 30 June 2021 from https://data.worldbank.org/indicator/NY.GDP.MKTP.KD.ZG?locations=CA.

World Bank (2021d) 'GDP Per Capita (Constant 2010 US$)'. Retrieved 30 June 2021 from https://data.worldbank.org/indicator/NY.GDP.PCAP.KD?locations=CA.

World Bank (2021e) 'Fuel Exports (% of Merchandise Exports)'. Retrieved 15 July 2021 from https://data.worldbank.org/indicator/TX.VAL.FUEL.ZS.UN?locations=CA.

World Bank (2021f) 'Worldwide Governance Indicators 2019'. Retrieved 26 July 2021 from http://info.worldbank.org/governance/wgi/#reports.

World Bank (2021g) *Regulatory Indicators for Sustainable Energy (RISE)*. Retrieved 26 July 2021 from https://rise.esmap.org.

Warren, F.J. and Lemmen, D.S. (2014) 'Synthesis'. In: Warren, F.J. and Lemmen, D.S. (eds.) *Canada in a Changing Climate: Sector Perspectives on Impacts and Adaptation*. Ottawa: Government of Canada, pp. 1–18. Available at: https://www.nrcan.gc.ca/sites/www.nrcan.gc.ca/files/earth-sciences/pdf/assess/2014/pdf/Full-Report_Eng.pdf (accessed 8 July 2021).

WRI – World Resources Institute (2021) 'Climate Watch Canada: Greenhouse Gas Emissions and Emissions Targets'. Retrieved 17 July 2021 from https://www.climatewatchdata.org/countries/CAN.

8 Conclusions

Challenges and opportunities for EU foreign policy and its analysis in an era of decarbonisation

This study investigated the implications that the phase-out of fossil fuel use and broader decarbonisation processes may have for European Union (EU) external relations with fossil fuel exporters. As trade in oil, gas, and coal – and other high-carbon products such as palm oil – accounts for a sizeable proportion of EU trade with many of these countries, the decarbonisation of the European and global economy is set to have important repercussions for bilateral relations. Our study has examined these repercussions in more detail with respect to a cross-section of six fossil fuel exporting countries from different world regions covering different resource endowments and stages of development, beyond the 'usual suspects' like Russia, Norway, and Algeria. Beyond assessing the importance of fossil fuels and their export for Azerbaijan, Canada, Colombia, Indonesia, Nigeria, and Qatar, the study paid particular attention to the potential of these countries and their bilateral relations with the EU to productively move beyond fossil fuels and enter a new level of cooperation. In this respect, it in particular also tried to identify how EU foreign policy could help develop the relations with these countries beyond fossil fuels.

This concluding chapter synthesises the main findings of the country studies, discusses the study's overall contribution to advancing the state of the art and identifies next steps in addressing the unfolding research agenda on (EU) external relations under decarbonisation. While our country selection does not necessarily ensure full representativeness, the case studies offer insights into a considerable range of different conditions. In the following, we highlight common features across the countries as well as the range that exists within our sample. We identify five key findings that relate to (1) the political and economic challenge of decarbonisation for the studied fossil fuel exporters, (2) the fragility and security context of these countries, (3) their climate policies and politics, (4) existing cooperative arrangements with the EU, and, last but not least, (5) the scope and starting points for the EU to fruitfully develop relations with these six countries under decarbonisation. Subsequently, we discuss the contribution of the study and its findings to key literatures (as identified in Chapter 1). The chapter concludes with a brief discussion of avenues for future research.

Key finding I: decarbonisation constitutes a key political and economic challenge for fossil fuel exporters

The case studies demonstrate that the decarbonisation imperative constitutes a key political and economic challenge for exporters of fossil fuels and other carbon-intensive

DOI: 10.4324/9781003183037-8

products. The economies and government budgets of these countries are, with some variations, highly dependent on the production and export of these goods. As a result, and again to varying degrees, they are also exposed to fluctuations in the price of fossil fuels, and continuing investments in related infrastructure carry a high risk of creating stranded assets.

High dependence on fossil fuel production and export and exposure to price fluctuations

The case studies illustrate that the economies and government budgets of a number of fossil fuel exporting countries are highly or significantly dependent on the production and export of oil, gas, coal, and/or other carbon-intensive goods. While for all of them these goods are a very significant part of their economy and related revenues are a significant contributor to public budgets, for some of them they are the main and even all-dominant sector, and/or their public budgets are overwhelmingly dependent on income from these dominant sectors.

Fossil fuel production and export account for significant shares of these countries' economies, ranging from less than 2 percent to about 25 percent of gross domestic product (GDP). In Azerbaijan, the oil and gas sector accounted for about 25 percent of GDP in 2019, while in Qatar it contributed one-fifth of GDP. While the share tends to be smaller in bigger countries and economies, it is still significant with almost 10 percent in Nigeria and close to 2 percent in Canada. Colombia's fossil fuel sector is relatively limited but sizeable at about 4.3 percent of GDP in 2019. In the case of Indonesia, fossil fuels and palm oil contribute an estimated total of around 7 percent of GDP.

Fossil fuels account for even bigger shares of these countries' merchandise exports (see Figure 8.1 below). In 2019, fuel exports accounted for over or close to 90 percent of merchandise exports from Azerbaijan, Nigeria, and Qatar, illustrating extremely high levels of export earnings from fossil fuels and the significance of these goods for trade relations in general. Nearly half of Colombian export earnings in 2019 can be attributed to fossil fuels (45 percent). In Indonesia, fuels provided for about 20 percent of Indonesia's foreign exchange earnings in 2017, whereas agricultural products (including palm oil) made up nearly 30 percent of exports. In 2019, palm oil accounted for more than 16 percent of Indonesia's foreign exchange earnings from merchandise trade. In 2019, fuels accounted for about 24 percent of Canadian goods exports.

Fossil fuels and fossil fuel exports, accordingly, also make major contributions to government budgets. These contributions may come from direct foreign exchange earnings, royalties, and taxes. The oil and gas industries have consistently accounted for more than 50 percent of Qatar's state budget and provided the largest share of income for the state budgets of Azerbaijan and Nigeria (fluctuating somewhat with the oil price). Fossil fuels have contributed less than 10 percent to government budgets in Canada and Indonesia.

Fluctuations in the price of fossil fuels in international markets illustrate the high dependence on fossil fuel production and export, as could be seen in the fall of the international oil price between 2014 and 2016 and is confirmed by the dramatic falls resulting from the Covid-19 pandemic in 2020. These falls have had a major impact on the economic development and public budgets of the six countries studied, in some cases causing recessions (with the Covid-19 impacts still unfolding). Although decarbonisation has not

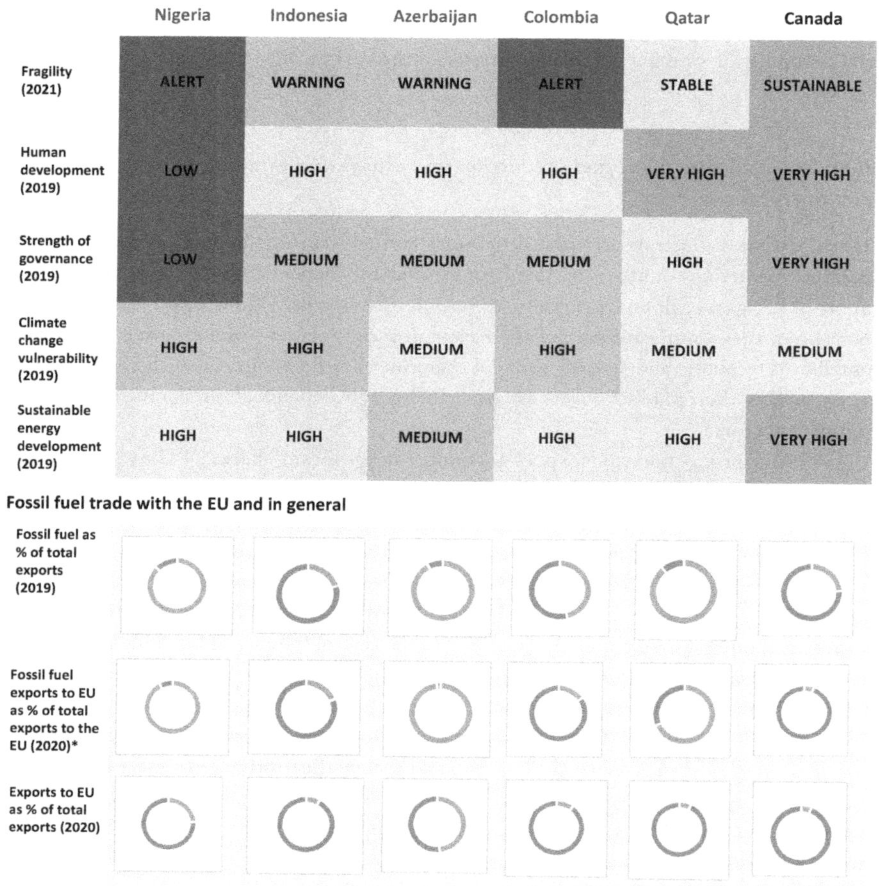

Figure 8.1 Governance indicators and fossil fuel exports for Nigeria, Indonesia, Azerbaijan, Colombia, Qatar, and Canada.

Sources: Governance indicators based on: Fund for Peace (2021), ND-GAIN (2020), UNDP (2020), World Bank (2021a, 2021b); trade data based on: European Commission (2021a, 2021b, 2021c, 2021d, 2021e, 2021f); https://data.worldbank.org (fossil fuel as a percentage of merchandise export; oil, gas, and coal as a percentage of GDP).

* In the case of Indonesia, palm oil export to the EU.

been a major driver of international fossil fuel markets to date, it could reinforce downward pressure on fossil fuel prices in future.

The drop in international oil and coal prices from 2014 to 2016 left a clear mark on the economies of the fossil fuel exporters investigated, in some cases leading to economic crises. For example, Nigeria experienced its first full year of recession in 25 years in 2015, with real GDP contracting by 1.6 percent. In Azerbaijan, GDP nearly halved from

2014 to 2016, and in Qatar it contracted by more than one-fourth (all in current US$). In Colombia, GDP growth fell from 4.9 percent in 2013 to 2 percent in 2016, and the contribution of fuel exports to GDP decreased from 11 percent in 2013 to 6 percent in 2016. Canada also experienced a decline of the growth rate from 2.9 percent in 2014 to 1 percent in 2015. In Indonesia, the economic impact was somewhat balanced by a decrease of fossil fuel subsidies that was facilitated by the price drop. In most cases, the economic repercussions of the Covid-19 pandemic, which led to a sharp decline in fossil fuel prices and further-reaching economic crisis, even exceeded those of the fossil fuel price crisis of 2014/2015.

The price drop also had major knock-on effects for public budgets. In Nigeria, public debt rose from 17.6 percent of GDP in 2012 to 29.1 percent in 2019. Azerbaijan and Qatar experienced even greater increases in public debt between 2014 and 2016, from 7.3 percent in 2008 to more than 22 percent in 2016 and from 32.3 percent in 2014 to 71.8 percent in 2020, respectively. Colombia experienced a decrease of the extractives sector's contribution to national fiscal income from 19 percent in 2013 to 5 percent in 2016. Canada saw its public debt ratio increase from 85.7 percent of GDP in 2014 to nearly 92 percent of GDP in 2016.

Although the economies and public budgets of fossil fuel exporters were all hit by the fall of fossil fuel prices (depending on the overall dependence on fossil fuel exports), the effects also depended on the instruments available to cushion the impacts. For example, the State Oil Fund of Azerbaijan was created to support the development of the country's non-oil sectors, but it has in reality served to balance price and resulting revenue fluctuations. In a more ad hoc fashion, the Indonesian government has managed the effects of price decreases to some extent by reducing fossil fuel subsidies. Canada has generally been less vulnerable to price fluctuations due to the advanced diversification and development of its overall economy, including non-oil sectors.

In general, the still-evolving Covid-19 crisis has largely confirmed and reinforced these findings. As a result of the steep fall in demand, prices of oil and gas also decreased, and fossil fuel exporters saw their exports and export earnings decline. This in turn put pressure on public finances and led to an increase of public debt. It also led to a broader reduction of GDP. Overall, the Covid-19 crisis therefore once more illustrated the dependence of the selected countries on fossil fuel production and exports and their vulnerability to price and demand fluctuations (even though the Covid-19 crisis admittedly also affected various other commodity markets).

High risk of stranded assets and insufficient diversification

The countries investigated have continued to invest heavily in fossil fuels and related high-carbon infrastructure, entailing a high risk of stranded assets under decarbonisation. This has contrasted with lower investment in non-fossil fuel sectors and at times even undermined progress towards economic diversification.

Fossil fuel assets at risk from stranding range from as yet untapped hydrocarbon reserves to fossil fuel sector infrastructure. In all the countries studied, proven reserves could support fossil fuel production well beyond 2050. For example, proven reserves could support current Nigerian oil production for almost another 50 years and current

Indonesian gas and coal production for more than 20 and 60 years, respectively. Azerbaijan's oil reserves are forecast to last for another 20 to 30 years, and the country could maintain current levels of gas production until 2090. Colombia could maintain current levels of coal production until the 2070s, and Canadian oil and coal reserves and Qatari gas reserves would support current production levels for more than 100 years. Important carbon-intensive infrastructure highlighted in the case studies includes oil and gas extraction facilities, networks of pipelines and other fossil fuel transport infrastructure (especially in Qatar, Azerbaijan, Canada), and fossil fuel-based power stations (especially coal, Indonesia). In Qatar, US$12 billion in investments in the oil and gas sector were announced in 2018, out of a total of new investments of US$85 billion in different sectors. Canada has also continued to make considerable investments into fossil fuels (especially oil sands). In 2018, total Canadian energy assets amounted to CAN$685 billion. The major share of foreign direct investment totalling nearly US$15 billion in Azerbaijan in 2017 went to oil and gas.

Progress towards economic diversification has been varied but has generally remained insufficient/slow, also as a result of a continuing focus on fossil fuels. The aforementioned State Oil Fund of Azerbaijan has aimed to support economic diversification – with very limited results. The Qatar Investment Authority set up in 2005 has a similar purpose but has in fact diversified to some extent the fossil fuel sector itself (from oil to gas) and has still left the overall economy hugely dependent on oil and gas. Efforts at economic diversification in Nigeria have also faced major obstacles, including an appreciating exchange rate due to the oil-based foreign currency earnings (the so-called Dutch disease), a neglect of other key economic sectors (such as agriculture), and a high dependence of industry and services on imports of inputs and raw materials.

The Canadian and Colombian economies are more diversified, but particular regions within these countries are heavily dependent on fossil fuel production and face particular challenges in transitioning towards decarbonisation. In Colombia, coal production is concentrated in the Departments of Cesar and La Guajira where it accounted for around 40 percent of GDP and 30,000 direct jobs (2015). In Canada, while overall the fossil fuel industry only accounts for a smaller part of the economy, it is of greater importance for the provinces of Alberta, Saskatchewan, and, partially, British Columbia.

Key finding II: the decarbonisation challenge frequently intersects with various other fragility and security risks

The decarbonisation challenge in fossil fuel exporting countries frequently intersects with other fragility and security risks, including conflicts at national and regional levels; weak and fragile governance arrangements; and impacts of climate change. Such risks and their severity vary across countries. These risks can exacerbate the difficulty of moving away from fossil fuels, and decarbonisation could in turn enhance these risks if not addressed adequately.

National and regional security risks

Our case studies illustrate that fossil fuel exporting countries frequently face significant national and/or regional security risks. Indeed, all our case study countries except Canada have encountered substantial security challenges; these take different forms and are highly specific to the respective contexts.

For Colombia and Indonesia, the main security challenges are domestic. Indonesia has a long history of separatist movements and clashes between divergent ethnic and religious groups (as well as discontent over unequal treatment of indigenous groups). In addition, terrorism remains a palpable threat to security. Colombia has faced longstanding internal conflicts with insurgent groups, in particular the Revolutionary Armed Forces of Colombia (FARC). After the 2016 peace agreement, the situation has remained fragile, and the country is in the midst of complex and nonlinear peacebuilding and post-conflict development processes. In addition, the Venezuelan political crisis that escalated in 2018–2019 has affected Colombia, which accommodated more than 1 million Venezuelan refugees.

In the cases of Qatar and Azerbaijan, security risks relate in particular to the regional context. Being located in the Middle East and the Gulf region, Qatar is part of a sensitive and challenging security environment, including the rivalry between Iran and Saudi Arabia for regional dominance and the involvement of various external actors (the United States, Russia, and Turkey in particular). In this context, deteriorating relations with Saudi Arabia and other Gulf countries in the 2010s culminated in them launching an embargo against Qatar from 2017 to early 2021 – which also left the Gulf Cooperation Council in paralysis for several years. With regard to Azerbaijan, the country is in a conflict with Armenia over the enclave of Nagorno-Karabakh, which is entirely situated in Azerbaijan but has an Armenian ethnic majority. After the conflict had been 'frozen' for many years, it escalated into a full-fledged war in the second half of 2020, and tensions remain high even after a Russian-mediated ceasefire was reached in November 2020.

Nigeria faces both serious internal and regional security threats, which pose a risk to political and economic stability. The security challenges include longstanding secessionist sentiments and movements in the southeast (in the wake of the secessionist Biafran War from 1967 to 1970) and militant groups in the Niger Delta seeking control and revenues from the oil industry. Further to the north, the insurgency of the Islamist sect Boko Haram has constituted a serious security risk since 2012. It also connects to the broader and growing security threat posed by Islamist extremists across several countries in the Sahel region (including Mali, Niger, and others). Many of the security challenges have their roots in acute socioeconomic inequality and conflicts, for example intensifying competition for fertile land and control over oil resources.

Under these circumstances, decarbonisation can hamper efforts to alleviate domestic conflict, while also creating opportunities. In particular, declining fossil fuel exports could threaten economic stability and government spending commitments essential for conflict management and post-conflict development. In Colombia, for example, the long-term success of the peace agreement with the FARC requires the government to honour its commitments to major investments and reform in rural areas. The resulting significant budgetary pressures Colombia has faced in recent years have already been aggravated by fluctuations in its key commodity markets and demonstrate the challenge arising from decarbonisation. As another example, volatility in oil markets may place further strain on Nigeria's economy and exacerbate other interrelated fragility risks, such as major gaps in public infrastructure, high inequality, rapid population growth and urbanisation, and growing insecurity of rural livelihoods in its Middle Belt and northern states. Having said that, advancing decarbonisation policies can also create opportunities that provide leeway to counter negative effects (e.g., through the expansion of renewable energies).

Fragile governance contexts

Most of the fossil fuel exporting countries also face governance challenges at the national level, although their significance varies (see also Figure 8.1). These seriously curtail the respective government's capacities to effectively address dependence on fossil fuel exports, and, more generally, to implement targeted policies to foster low-carbon development. The main exception is Canada, which is a developed and mature parliamentary democracy (with longstanding tensions between its French-speaking and English-speaking regions having been successfully managed in the twenty-first century).

Nigeria stands out as being particularly debilitated in this respect. While characterised by increasing inequality and significant demographic pressures, Nigeria also suffers from high levels of corruption and weak governance. The ability of the state to collect taxes and control its external borders (and internal territory) is therefore seriously curtailed. Key indexes point to low levels of human development and poor governance, while the Fragile States Index places Nigeria in the 'alert' category (see Figure 8.1).

Indonesia and Azerbaijan also face significant challenges. In addition to grappling with high levels of inequality, Indonesia faces issues regarding discrimination and state capacity. While strength of governance is rated as medium, the Fragile States Index remains at 'warning' level. Azerbaijan is characterised by serious issues of corruption and political freedom. The 2021 Freedom House Index categorises Azerbaijan as 'not free' and curtailment of political rights, repression of civil liberties, and stifling of press freedom are identified as shortcomings.

Qatar and Colombia face fewer governance challenges. Qatar has low levels of corruption or repression but faces challenges with respect to democracy and political rights (low-level general freedom, civil liberties, political participation). A particular issue is the problematic situation concerning around 2 million expatriate workers, mainly low-skilled Asian construction workers. Qatar has been criticised for their exploitation and for the lack of integration of these Asian workers into the society. In reverse, Colombia has developed relatively strong democratic institutions and has improved on the rule of law, although its recent rating in the Fragile States Index has worsened and corruption continues to constitute a central challenge (along with high levels of inequality). Overall, governance challenges present less of an impediment to addressing fossil fuel dependency.

Climate change impacts

Climate change is predicted to have a medium to high impact on the countries studied. These impacts could strengthen support for engaging in climate policy and related cooperation. However, the concrete impacts might also restrict the capacity to act on fossil fuel dependency, as priority might be given to urgent adaptation measures.

Climate change vulnerability is assessed as medium in Canada, Azerbaijan, and Qatar (Figure 8.1). Temperatures in Canada have already increased more than 1.5°C, approximately double the global average and significant impacts are particularly being felt in coastal and northern parts of the country. At the same time, Canada may also see some positive impacts, such as growth in agricultural food production in northern regions. The country also scores relatively favourably on indexes measuring vulnerability and readiness. Qatar is affected by sea-level rise (with 96 percent of its population living in coastal towns and cities and much of its infrastructure in peril) and the associated risk of groundwater salinisation. Further increases to already high temperatures and aggravated water scarcity also constitute significant challenges, while the country scores slightly above average on readiness. In Azerbaijan, rising temperatures constitute a threat for the agricultural sector due to resulting increased evaporation, droughts, and water shortages – while the sea level of the Caspian Sea may in fact fall as a result of increased evaporation. The country scores medium on vulnerability and readiness.

Climate change vulnerability is considered high in Colombia, Indonesia, and Nigeria. Colombia is subject to a variety of climate change impacts, including sea-level rise, more frequent and more intense La Niña and El Niño phenomena, and decreased rainfall. Higher temperatures and increased water scarcity (due to decreasing run-off from the mountains) are of particular concern in the Andean region, home to 75 percent of the Colombian population. With the second longest coastline in the world, Indonesia is particularly impacted by changing rainfall patterns, temperature and sea-level rise, and extreme weather events and natural disasters. Jakarta is the world's fastest-sinking city, with around 40 percent of the city below the sea level. Nigeria is already suffering from increasing temperatures, erratic rainfall, desertification, rising sea levels, and drought. This puts further pressure on an agricultural sector that is already struggling to maintain or enhance productivity, as well as urban centres, not least Lagos.

Key finding III: climate policy frameworks are in urgent need of further development, but ambitious climate policies face significant resistance

While climate policy frameworks are at various stages of underdevelopment, progress in the development of adequate targets and policies of the studied countries is hampered by significant socioeconomic barriers, in particular strong opposition from the fossil fuel industry that is frequently closely related to governments.

Climate policy frameworks at various stages of underdevelopment

The development of climate policy frameworks in the studied fossil fuel exporting countries has remained inadequate. All six countries are parties to the Paris Agreement and consequently have 'Nationally Determined Contributions' (NDCs). All except Qatar have included economy-wide greenhouse gas (GHG) emission reduction targets for 2030 in their NDCs. As an industrialised country, Canada aims to reduce its emissions by 40–45 percent below 2005 levels, while Azerbaijan's target of keeping emissions 35 percent below 1990 levels can be considered relatively ambitious (but lacks implementing policies). Nigeria, Indonesia, and Colombia aim to reduce emissions by 20 percent (Nigeria), 29 percent (Indonesia), or 51 percent (Colombia), compared to projected business-as-usual scenarios and establish higher targets on the condition that sufficient international assistance is made available. Qatar's NDC does not establish any emission target.

While these targets are generally insufficient for achieving the objectives of the Paris Agreement, the (under)development of domestic policy frameworks also raises concerns about the capacity to effectively implement them (and any upgraded targets in the future). Within the sample, Canada and Colombia are the most advanced in developing their domestic climate policies. Canada has advanced its domestic policy frameworks, especially since the Paris Agreement, with a national carbon pricing mechanism at its core and a number of other measures addressing various sectors. While this framework still needs to be further strengthened to meet Canada's enhanced emission reduction target and to advance towards climate neutrality, its long-term stability is in question given that climate policy has been a highly partisan issue. Changes of government at both federal and provincial levels have therefore hindered the development of a consistent trajectory. Colombia has also developed a relatively robust institutional, legal, and policy framework for implementing its NDC (that will, however, require further elaboration and strengthening) and advancing both climate mitigation and adaptation. It has actively promoted market mechanisms and introduced a carbon tax in 2017. The country's first climate change law was passed in 2018.

In contrast, effective climate policy frameworks in Indonesia, Nigeria, Azerbaijan, and Qatar are at best at an early development stage. Indonesia has elaborated general action plans on climate change mitigation and adaptation, including a moratorium related to deforestation, its main source of GHG emissions. However, progress on policy measures for addressing GHG emissions in other growing sectors has lagged. Similarly, Nigeria has established general frameworks for addressing mitigation and adaptation, but concrete measures need further elaboration and implementation. While Azerbaijan has reported the adoption of a number of relevant legislative acts and established supporting institutional structures, climate change policy, incentives, or regulations have nevertheless been found to be absent. Finally, Qatar does not have a firm framework for its domestic climate policy but appears to focus on developing its economy, reaping co-benefits for GHG mitigation where opportune (e.g., some investments in renewable energy).

Significant opposition to transformational climate action

The underdevelopment of domestic climate policy frameworks in the studied countries correlates with significant politico-economic opposition to transformational climate action. Even in Canada and Colombia where fossil fuels have a comparatively modest importance for the national economy, this opposition has been significant and strong. In Canada, climate policy has remained a very partisan issue, with conservatives opposing and undermining climate action. Opposition is particularly strong in the fossil fuel-rich provinces (Saskatchewan, Alberta, and British Columbia), which hold significant political power in the Canadian federal system. As a result, oil and gas pipelines and exploitation of oil sands have also received support from liberal governments. While climate change has been less contested as a political issue in Colombia, support for fossil fuel extraction and mining activities, while slowly receding, has remained strongly embedded in the politico-institutional system. As in the case of Canada, confronting fossil fuel interests has been complicated by strong dependence of particular regions on fossil fuel exploitation. At the same time, support for addressing climate change has also been growing in both Colombia and Canada.

Fossil fuel interests have been even more strongly entrenched in the other four countries studied. Most importantly, fossil fuel extraction has been either dominated by state-owned companies (Qatar, Azerbaijan) or such enterprises have at least played an important role in the sector (Nigeria, Indonesia). As a result, the formal and/or informal links between fossil fuel industries and the government have generally been very close, at times also in the context of corruption. At the same time, societal and political support for decisive climate action has remained too weak to balance strong fossil fuel/high-carbon interests across these countries.

Key finding IV: EU external relations can build on various pre-existing cooperative arrangements

Existing cooperative arrangements between the studied fossil fuel exporters and the EU provide a sound basis for developing the bilateral relations fruitfully beyond high-carbon products. The most important institutional arrangements in this respect include Partnership and Cooperation Agreements and Free Trade Agreements, but existing cooperation is also based on various regional and multilateral forums.

Partnership and Cooperation Agreements, high-level political dialogues, and Free Trade Agreements

EU relations with the studied countries can build on existing Partnership and Cooperation Agreements or other forms of high-level dialogues. Building on decades of close political cooperation and consultation, the EU and Canada launched a Strategic Partnership Agreement, alongside the Comprehensive Economic and Trade Agreement (CETA), in 2016, and they hold annual high-level bilateral summits. The EU and Azerbaijan have cooperated under the umbrella of a Partnership and Cooperation Agreement since 1999 (with an update under negotiation since 2017). EU-Indonesia dialogue has a firm basis in the Partnership and Cooperation Agreement of 2014, and the EU and Qatar have strengthened cooperation under a Cooperation Agreement concluded in 2018. Without a formal overarching agreement, the EU and Colombia have engaged in bilateral and high-level political dialogues and have explored new mechanisms for 'lifting bilateral relations to a new level' (EEAS 2019). While primarily rooted in the Cotonou Agreement with African, Caribbean, and Pacific developing countries (and its successor agreed in late 2020), EU-Nigeria relations have also been developed under a political framework for dialogue, the 'Nigeria-EU Joint Way Forward'.

In several cases, bilateral trade relations find a further firm basis in relevant free trade agreements. Perhaps the best-known among them is the aforementioned CETA between Canada and the EU. Trade between Colombia and the EU is governed by a multiparty free trade agreement formalised in 2013. Negotiations on a Comprehensive Economic Partnership Agreement between the EU and Indonesia have been ongoing since 2016. No free trade agreements are currently in sight with Nigeria (which has rejected joining the Economic Partnership Agreement between the EU and the Economic Community of West African States), Qatar (with negotiations between the EU and the Gulf Cooperation Council countries suspended in 2008), and Azerbaijan (with trade to be covered in the broader Cooperation Agreement under negotiation – see above).

Nevertheless, the trade and investment relationship of all studied countries with the EU is very substantial. In 2019, EU exports accounted for more than 35 percent of Qatari and more than 20 percent of Azeri imports, whereas EU imports made up more than 10 percent of Qatar's overall exports and about half for Azerbaijan. Canada is the EU's 11th top trading partner, representing about 1.5 percent of total EU external trade, while the EU is Canada's and also Colombia's third largest trading partner after the United States and China. Nigeria is the destination for around half of EU export to the West African region and the origin of close to 70 percent of the imports. Also, the EU accounted for 5.5 percent of Indonesian commodities exports. The EU is also a significant investor in Colombia. The stock of EU-Canada direct investments amounted to some €640 billion in 2019. Qatar has also become a significant investor in the EU after the financial crisis, with Qatari investments in the EU surpassing EU investments in Qatar by a factor of three (2016). The importance of fossil fuel trade between the studied countries and the EU is clear from Figure 8.1.

Other fields and institutions of cooperation

Beyond these formal agreements, the studied fossil fuel exporters and the EU also cooperate within a web of other bilateral, regional, and multilateral fora. How tight this web is, and where its focal points lie, varies between the countries, but it generally provides a solid basis for developing cooperation. This web of cooperation is perhaps closest with Canada, as the EU (and its member states) and Canada actively collaborate in many forums, including NATO, the UN, the G7, the G20, the WTO, the OECD, the IAEA, and others. Since 2017, Canada, the EU, and China have convened a Ministerial on Climate Action, and Canada and several EU member states are members of the Powering Past Coal Alliance. Bilateral scientific and technological cooperation has been developed under a targeted Agreement for Scientific and Technological Cooperation since 1996.

However, cooperation with the other studied countries also has a much broader basis than fossil fuels, shaped by the particular conditions in each country. The EU and Colombia have cooperated closely to set Colombia on the road to peace and have also sought to collaborate in a number of political and policy consultations. This includes the area of security and defence under a Framework Participation Agreement on the participation of Colombia in EU crisis management operations. Relevant regional fora include the Community of Latin American and Caribbean States (covering various issues, including research and innovation, education, migration, etc.) and regional policy dialogues on economic, social, and territorial cohesion. The EU and Colombia have also been close partners in the United Nations Framework Convention on Climate Change (UNFCCC) and international climate policy more generally.

EU-Azerbaijan cooperation finds another firm basis in the European Neighbourhood Policy and the Eastern Partnership. Azerbaijan also is a part of the EU Erasmus+

programme, and there is a bilateral Mobility Partnership (signed in 2013). The EU has also taken a role in managing the conflict with Armenia over Nagorno-Karabakh (although not most prominent). With Nigeria, there is established cooperation on security and counterterrorism, humanitarian aid, and rural and economic development (including investment). The aforementioned Cotonou Agreement (to be succeeded by a new Post-Cotonou Agreement) has provided a strong multilateral basis. EU-Qatar cooperation has been strongly embedded in EU cooperation with the Gulf Cooperation Council and thus suffered due to increased tensions between Council's member states in the 2010s. The EU offered to mediate in this conflict and has otherwise focused on developing socioeconomic aspects of the relationship. EU-Indonesia relations are embedded in the Association of Southeast Asian Nations (ASEAN). Both sides concluded a Voluntary Partnership Agreement on legal timber exports to the EU in 2013, and the EU and its member states are proactive in a number of cooperation projects, including some to tackle climate change and deforestation. Since the late 2010s, cooperation has been under strain in the wake of the EU's decision to phase out the recognition of palm oil-based biofuels under its Renewable Energy Directive.

Key finding V: there is ample potential for developing EU external relations with fossil fuel exporters beyond fossil fuels

Furthermore, our case studies indicate that there is a strong and varied basis for fruitfully developing EU external relations beyond fossil fuels even with those countries that may be considered particularly hard cases, namely fossil fuel exporters. Partners highly dependent on the production and export of coal, oil, gas, and other high-carbon products also have other significant interests that provide entry points for developing climate-neutral EU foreign relations. As mentioned in Chapter 1 and illustrated in Figure 8.2, we focus on five prime areas to describe and assess the room for advancing decarbonised EU external relations: (1) climate and energy, (2) trade and investment, (3) science and education, (4) finance and development, and (5) security and peace.

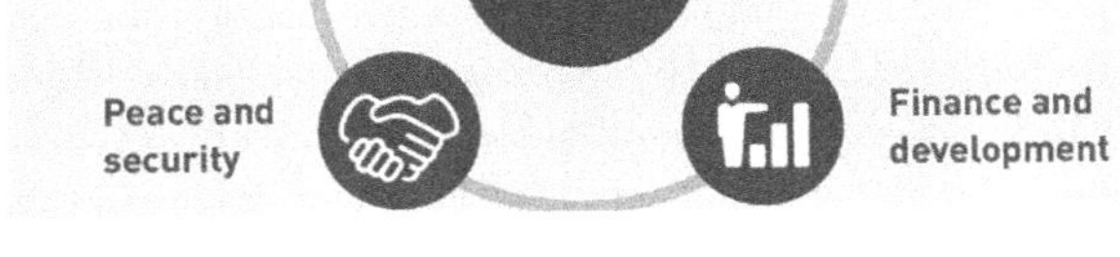

Figure 8.2 Five areas for building EU external relations under decarbonisation.
Source: Authors.

Climate and energy

Low-carbon development, including renewable and clean energy technologies, is a particularly promising field for developing future cooperation because there is considerable scope for stepping up action worldwide and the EU is relatively advanced in its climate and energy transition by international comparison. With countries such as Canada, Colombia, and Indonesia, the EU shares challenges related to a 'just transition' for regions particularly dependent on high-carbon assets, including fossil fuel production, and this can provide suitable levers for closer cooperation. Furthermore, enormous potential to further intensify cooperation on renewable energy exists in virtually all studied countries. Developing renewable energy can, in a first step, enhance availability of oil and gas for export (especially in Azerbaijan and Qatar). Beyond that, renewable energy has increasing price advantages, is becoming an internationally recognised prime energy resource, and can be linked to established programmes and strategies for economic diversification. While taking into account country-specific circumstances (e.g., high solar potential in Qatar and Azerbaijan, linkage with climate-smart agriculture and rural electrification in particular in Nigeria, general links with development finance for both Nigeria and Indonesia), cooperation can in several cases also put an emphasis on related (higher) education (see also below).

Climate and energy policies also provide a promising field of cooperation beyond renewable energy. Again, entry points vary between different countries in accordance with diverging interests and conditions. Shared interests in the development of climate policy with Canada could support cooperation on buildings, transport, and energy-intensive industries. Discussions on the future development of gas extraction and trade may be linked to the development of renewable energies and more ambitious climate policy especially regarding Azerbaijan and Qatar. Especially for the more vulnerable countries (Nigeria, Indonesia, Colombia), the exploration of the adaptation-mitigation nexus holds particular promise. And especially with Canada and Colombia, existing close cooperation on international climate policy in the UNFCCC and beyond can be further developed. As the interest in effective climate policies and measures continues to grow, targeted dialogues can explore the different aspects of this field.

Trade and investment

Trade, investment, and, more generally, economic cooperation with the studied countries have enormous further potential also beyond fossil fuels and other carbon-intensive products. The aforementioned partnership and cooperation agreements and free trade agreements provide a strong basis for such economic cooperation. Economic relations with Canada already go far beyond fossil fuels that only account for about 10 percent of Canadian exports to the EU. Similarly, EU-Colombia trade already goes much beyond fossil fuels, even if the latter still accounts for a significant share of Colombian exports to the EU; economic cooperation can in this case tie in with progressing Colombian plans for economic diversification, including a green growth strategy and the development of creative industries. Since palm oil and related biofuels have accounted for a significant share of Indonesian exports but are in questions due to EU concerns over emissions from indirect land-use change, sustainable biofuel production deserves particular attention in the evolving bilateral cooperation – along with sustainable forest management and sustainable urban development. In the cases of Nigeria, Azerbaijan, and Qatar, exports in general and to the EU have remained dominated by oil and gas. All three countries have longstanding strategies or plans for economic diversification that cooperation with the EU could more strongly build on and connect to (including regarding renewable energy – see above). This could also help enhance these countries' economic resilience in a world that is beginning to move beyond fossil fuels.

Science and education

Poorer fossil fuel exporters in particular face significant challenges in developing a knowledge-based economy, which is itself closely linked to aspirations for economic diversification. With considerable variation, education and skills development remain a particular point of attention for most countries. While Canada has a developed educational system, Azerbaijan faces significant challenges in this respect. Despite sizeable progress in the 2010s (also as part of the Azeri strategy for economic diversification), the supply of skilled graduates does not suffice to meet the demand and Azerbaijan's workforce is considered medium-qualified. Qatar has invested significantly in education but still ranked 60th out of 77 countries in the 2018 Program for International Student Assessment (PISA). Despite targeted government efforts, related challenges in Nigeria have remained severe, and have even deteriorated in many respects, regarding both primary education and university training. Aspiring to become the most educated country in Latin America, Colombia has made good progress, but access to education varies significantly between socioeconomic groups. Finally, Indonesia has made huge progress in education in the twenty-first century, but still has some way to go to catch up with neighbouring countries such as Vietnam and Singapore.

Possessing advanced knowledge economies as well as established educational and research programmes, the EU and its member states have much to offer in cooperating with fossil fuel exporters on education and skills development as well as research. This prominently includes the EU Erasmus+ programme for educational exchanges and cooperation that several of the investigated countries are already tied into. Canada has also been an active participant in the EU framework programmes for research and Marie Sklodowska-Curie Actions, and the EU and Canada have even developed a bilateral Agreement for Scientific and Technological Cooperation. Building on such existing frameworks, cooperation with fossil fuel exporters on education, research, and training can be intensified bilaterally and/or in interregional frameworks. Renewable energy and other aspects of a low-carbon economy hold a particular potential in this respect (e.g., related university partnerships or joint university degrees).

Finance and development

The EU already has strong finance and development cooperation with many of the countries studied. As a part of the European Neighbourhood Policy, Azerbaijan receives support from and has had access to a number of financial and support mechanisms, including the European Neighbourhood Instrument. The EU and its member states have also worked with Indonesia on a range of relevant initiatives, including for achieving the Sustainable Development Goals (SDGs) and addressing climate change (as well as in the context of ASEAN). Furthermore, development assistance to Nigeria has been significant and has covered energy, low-carbon development, and climate change adaptation, in addition to humanitarian aid, health, governance, and other areas. The finance and development relationships with Colombia and Qatar (and Canada) are based less on bilateral aid (even though Colombia has received both bilateral and multilateral climate finance support) and rather are characterised by significant investment flows. For example, the EU is a sizeable investor in Colombia, and Qatari investments in the EU have been significant.

Significant scope remains for realigning finance and development cooperation to support decarbonisation. This can take inspiration from general debates on green and climate finance within the EU, for example, on the EU budget, the sustainable finance taxonomy, and guidelines of the European Investment Bank (e.g., Kavvadia 2021; Ohnesorge and Rogge 2021). Accordingly, external finance and development cooperation could aim (1) to ensure a significant share of overall finance (at least 30 percent) is reserved for climate and low-carbon development purposes and (2) to phase out or prohibit finance that is not aligned with low-carbon development objectives. An important element of such a reformed policy on external finance and cooperation would be resources for the support of a just transition that would heed the lessons from EU internal debates on the need for support for regions particularly dependent on high-carbon industries and activities. For example, the EU could provide assistance to set up just transition task forces to engage directly with stakeholders in affected regions and mobilise political and financial support for the development of new employment opportunities there. Such approaches have been tried and tested in both the EU and Canada.

Security and peace

Decarbonisation is also closely related to issues of peace and security. As mentioned above, all fossil fuel exporters studied except Canada have faced serious internal and/or external security challenges (that have been reinforced, to varying extents, by weaknesses of governance systems and worsening climate change impacts). While the main security challenges are domestic for Colombia and Indonesia, Qatar and Azerbaijan are embedded in precarious regional security contexts. Nigeria faces both serious internal and regional security threats. These security challenges, while tending to exacerbate efforts at low-carbon development, provide an important field of cooperation for the EU.

To this end, the EU can build on existing cooperation with fossil fuel exporters on these matters. For example, the EU offered its services to mediate in the conflict between Qatar and partners in the Gulf Cooperation Council. It has also promoted confidence and peace-building activities in the conflict between Azerbaijan and Armenia and has supported the rule of law and the fight against corruption in Azerbaijan. It has also actively supported the peace process in Colombia and has closely cooperated with Nigeria on addressing the humanitarian and security crisis, as well as on strengthening good governance.

Whereas we have discussed the above five policy areas for developing EU external relations beyond fossil fuels separately, they are interrelated. As such, it is important to ensure coherence across these areas of cooperation. For example, cooperation on climate and energy will not least be fed by channelling finance and development co-operation accordingly. Similarly, and as indicated above, cooperation on science and education can reinforce cooperation in other areas. Cooperation in the different areas may hence usefully be pursued in an integrated manner through long-term engagement strategies that consider interactions and maximise coherence and synergy across the five building blocks. Successfully developing relations with partner countries in a decarbonising world requires the EU to take an active and targeted approach. Fruitful external relations are unlikely to come about by themselves but need to be shaped via an active foreign-relations strategy. Long-term engagement strategies with individual countries and regions could form an effective and concrete means for enhancing sustainability across EU external relations in line with the European Green Deal (European Commission 2019).

Beyond the state of the art

How do the findings of this study advance the scientific state of the art? In the introductory chapter, we have built on four strands of literature on (1) EU climate diplomacy and foreign climate policy; (2) EU foreign policy and external energy relations; (3) environment/climate and security, and (4) the geopolitics of energy under decarbonisation. Here, we discuss the contribution of our study to advancing each of them in turn.

EU climate diplomacy and foreign climate policy

As discussed in Chapter 1, the longstanding literature on EU climate diplomacy and foreign climate policy and governance has in particular investigated the EU external engagement focused on climate policy itself. It has hence largely focused on the role and effectiveness of the EU in affecting *climate* policy developments beyond its borders, be it at the multilateral, transnational, regional, national, or subnational levels. Much of this literature has especially explored the EU international 'leadership' on climate change (e.g., Wurzel et al. 2017; Adelle et al. 2018; Oberthür and Dupont 2021; Petri and Biedenkopf 2021). To the extent that bilateral relations have been the subject of this research, the literature has mainly investigated the EU climate relations with key countries such as China, India, the United States, Brazil, and others (Torney 2015; Adelle et al. 2018; Gurol and Starkmann 2021).

This study makes a twofold contribution to advancing this literature. First, it broadens the scope of the research interest beyond a focus on climate policy developments by also exploring the implications of the global and European decarbonisation process on the broader bilateral relations with other countries. We develop an approach for investigating the bilateral relations of the EU under decarbonisation more broadly – an approach that allows us to identify the challenges and opportunities for the EU to advance both climate policy in bilateral relations and the bilateral relations more generally. The investigation of EU climate diplomacy thereby becomes embedded in the larger context of general bilateral relations, which facilitates the recognition of linkages across different policy fields. As a result, each of our case studies could identify scope for fruitfully developing EU bilateral relations in climate and energy as well as in other policy fields, thereby enabling a more integrated approach (see above). While we are not arguing that a narrower focus on climate diplomacy is redundant, the emerging more integrated approach seems timely given the crosscutting nature of the political challenge of climate change – whose geopolitical significance is increasingly acknowledged, including in the European Green Deal (European Commission 2019; see also below).

Relatedly, the second major contribution of our study is to make a first step towards a much-warranted expansion of the investigation beyond the 'usual suspects', that is, the key countries and major emitters that have been in focus so far. From a climate perspective, the focus on fossil fuel exporters may be warranted because of the importance of addressing not only the demand but also the supply of fossil fuels for advancing the climate transition (Piggot et al. 2020; Rayner 2021). A broadening of the perspective beyond major emitters and economies becomes even more inescapable if we want to address the broader foreign policy challenges and implications that the climate transition entails, for example in the context of changing regional power dynamics. These challenges and implications transcend the major economies and emitters and call for a broad scope and perspective. In this vein, our focus on a selection of fossil fuel exporting countries constitutes just the beginning of a broadening research agenda that we discuss further below.

Overall, our broadening of the view on EU external relations under decarbonisation would appear to be in line with the evolving EU policy agenda as reflected in the European Green Deal (European Commission 2019). The Green Deal's external dimension aims at integrating climate and the environment into all external policies and relations. Consequently, our study can also be read as a contribution to the further development of the Green Deal's external dimension.

EU foreign policy and external energy relations

The literature on EU foreign policy and external energy relations has for a long time had a strong focus on issues of the security of energy supply and various related issues (relationship between the European Commission and member states, EU competence, EU-Russia relations, pipeline politics, type of EU action/strategy, etc.). Despite some relevant contributions, climate change and climate policy have overall not been at the centre of interest. If at all, they have mainly been taken into account as an intervening or contributing factor affecting to some extent the actual issue of interest (see Chapter 1).

This study has the potential to help move the climate transition more centre stage in the investigation of EU foreign (energy) relations in at least two ways. First of all, our findings substantiate the broader and more general relevance of the climate transition for EU partner countries and, consequently, bilateral relations. They illustrate that the political and economic challenges arising are momentous (Finding I) and that the climate transition intersects with other key fragility and security risks in important ways (Finding II). They also provide evidence for the difficult climate policy and politics context in the partner countries investigated (Finding III) and identify, beyond the challenges, also the opportunities and potential that decarbonisation may be able to foster and build on. Importantly, our study goes beyond advancing general propositions about these implications of the climate transition (as prevalent in current literature) by investigating, corroborating, and nuancing them 'on the ground', that is, with respect to specific countries and related conditions.

Furthermore, our analysis demonstrates how climate considerations and objectives are relevant and connected to, and can be integrated into, broader external relations. This already arises from the aforementioned focus on the exploration of links with other fragility and security risks. It gets into further focus in our investigation of the broader basis of bilateral relations between the EU and the partner countries (as addressed in Finding IV on existing cooperative arrangements), as well as the analysis of the potential for developing these bilateral relations fruitfully beyond fossil fuels (as synthesised in Finding V above). Again, the study's distinct contribution not least rests on its case-study approach that results in a wealth of concrete and nuanced substantiations of these points, which future studies can build on. On the whole, the overall study and its case studies provide a basis for advancing our understanding of the concrete implications of the European and global climate transition for bilateral relations and the scope and need for developing EU external relations.

Environment/climate and security

The scholarship on environment/climate and security has highlighted and investigated the role of environmental problems and climate change as threat multipliers. As discussed in Chapter 1, this literature has also more recently increasingly explored the effects of climate *policy* on the emergence (and mitigation) of conflict (e.g., Gemenne et al. 2014; Swatuk et al. 2021). A common defining feature of this line of research has been its special attention to the interaction of climate/environmental impacts and climate policy with other factors such as regions' fragility or other security risks.

Our study can be seen to make a twofold contribution to this line of research as well, which may also provide a fruitful basis for future research. To start with, the study pursues an approach that advances towards a more comprehensive appraisal of relevant factors and their interaction. Hence, our case studies explore climate change impacts, impacts of the climate transition, and other fragility and security risks at national and regional levels – and they pay specific attention to how the decarbonisation process may intersect with these risks (see Finding II above). Although itself not focusing on security and conflict, the study thereby nevertheless makes a contribution to framing and understanding the interaction of broader security and fragility risks with both climate change impacts and the effects of the climate transition. The latter effects may both exacerbate existing risks or help counter them (e.g., due to the new opportunities possibly created). Our study thereby also provides a valuable basis for future investigations focused on the varying security implications and their conditions (see also below on future research).

In addition, it again deserves highlighting that our case studies add valuable concreteness that illustrates and makes tangible the range of different factors and their constellations, including security risks. They thus provide demonstration material for how conditions for the interaction of these factors vary in the real world and result in differing outcomes. In our case selection, the range extends from Canada with low fragility and security risks to Colombia and Indonesia with primarily domestic fragility and security issues to Azerbaijan and Qatar with regional security challenges to Nigeria where security risks are both domestic and regional. This may well provide a basis for future more detailed and systematic explorations of the interaction between the climate transition and security (see also below).

The geopolitics of energy under decarbonisation

Finally, the evolving literature on the geopolitics of energy has generally begun to explore the effects of the global climate transition on the international distribution of power and interdependence against the backdrop of geographical factors and resource endowments. This has, among other things, included analysis of the decline of fossil fuels also central in this study (e.g., de Jong et al. 2017; IRENA 2019; Blondeel et al. 2021). As pointed out in Chapter 1, this literature has so far focused on broad and general effects across different groups and types of countries within their geographical contexts and has hence operated at a fairly aggregate level, taking a top-down perspective.

Based on concrete and detailed country case studies, our analysis makes a start towards matching this top-down perspective with a bottom-up approach. While the so-far-prevalent aggregate, top-down perspective has been valuable in teasing out main overall effects and trends, it has hardly been able to take into account the impacts of differing conditions across varying countries. As our case study-based findings illustrate, different contexts and constellations do matter and provide widely diverging conditions for the management and governance of the impacts of the decline of fossil fuels and the climate transition. They provide a good starting point to add nuance to, and complement, the existing aggregate perspective. At a minimum, the bottom-up case-study approach of our analysis can shed light on the more concrete underpinnings of the general impacts highlighted by the literature on the changing geopolitics of energy. In addition, this approach allows to trace the interaction with other factors that an aggregate perspective may not easily be able to capture. Upon systematic comparison and re-aggregation of the results of different case studies, this could also inform more nuanced and sophisticated insights at the aggregate level. Further case studies may be required for a firmer evidence-based foundation of such more general insights.

Finally, the explicit integration of a policy dimension in our analytical approach deserves highlighting. As a result of the aggregate perspective of the existing work on the changing geopolitics of energy (frequently rooted in an overarching political-economy approach), structural changes tend to be considered a given. By contrast, our bottom-up approach integrates policymaking in the partner countries (e.g., climate policy and policies towards economic diversification) and the EU (in particular external policies towards the partner countries) into the analysis as co-shaping the eventual impacts of the climate transition. While structural changes should be acknowledged and identified, proactive policymaking can shape (and mitigate) their impacts – rather than simply reacting to them.

Overall, the above discussion underscores two major crosscutting and distinctive contributions of our study to advancing the state of the art in the different literatures referred to. First, our case-study-based approach adds concreteness and nuance to the so far frequently more general and aggregate analyses of the implications of global and European decarbonisation on partner countries and the EU relations with them. Also, the group of countries in focus in our case studies (i.e., second-tier fossil fuel exporters) have so far hardly been in the centre of research attention. Second, our approach and research distinguish themselves by combining different perspectives, factors, and strands of literature. We attempt to take account of structural changes affecting EU bilateral relations with partner countries, while delimiting the scope for foresightful policymaking to fruitfully reshape these relations accordingly. The study does so while taking into consideration a variety of factors directly and indirectly affected by the climate transition as well as shaping its context. Finally, we thereby keep a double focus on fostering EU external relations *for* and *under* decarbonisation. The final section turns to possible avenues for future research to further advance this agenda.

Instead of a conclusion: avenues for future research

The study of the broader repercussions of European and global decarbonisation for foreign policy, external relations, and international relations in more general is still at an early stage. First steps have been made, including in the literatures discussed in the previous section, but many aspects warrant further elucidation. This study has attempted to shed some further light on a number of facets, in particular (1) the challenges and opportunities arising for second-tier fossil fuel exporting countries from the climate transition and (2) the scope for the EU to shape and reshape its relations with these countries in support of both the climate transition and, more generally, fruitful and productive bilateral relations. As an early contribution to a still emerging field of inquiry, our findings raise more questions for future research than they can themselves possibly answer. We here identify seven particular avenues for future research that arise from our study.

First, expanding the investigation to additional countries promises significant added value. Our study remained focused on six out of several dozens of fossil fuel exporting countries. Although we attempted to select them from the larger group so as to ensure variance of conditions, the examination of further countries of this group promises to create additional insights. The investigation of further countries and the EU relations with them should allow to confirm and further nuance and systematise the findings of our study. In addition, our findings indicate that each country has specific domestic and contextual conditions that need to be taken into account. Hence, defining strategies for developing future bilateral relations under decarbonisation will require country-specific analysis.

Second, in order to develop a more encompassing understanding of the challenges and opportunities of the European and global climate transition for EU external relations, the analysis should be expanded to other classes of countries as well as other transition challenges. Whereas our study focused on fossil fuel exporters, the climate transition will also, in yet specific ways, affect countries producing and exporting other high-carbon products such as palm oil (partially addressed in our case study on Indonesia), other agricultural and forest products (including timber), and products of energy-intensive industries (steel, chemicals, etc.). Beyond that, there are a number of other relevant developments implied by the climate transition implicating countries in differentiated ways, including the rise of hydrogen as an energy carrier, growing demand for certain raw materials (such as various rare earths), changes of international value chains as a result of more circularity, and so on. The exact shapes of these developments are in large part only emerging, and further changes may come to the fore as the climate transition advances. Consequently, future research may investigate relations with the implicated groups of countries.

Third, and following from the previous point, future research may usefully consider a fuller set of transition challenges in combination. While it is important to understand the implications of a fossil fuel phase-out for fossil fuel exporters (as in the current study) or the repercussions of the rise of 'green' hydrogen as an energy carrier for countries with related capacity, it is eventually the combination of these developments driven by the climate transition that needs to be understood and managed. For example, Australia and Gulf countries are implied both by the decline of fossil fuels and the rise of green hydrogen (and renewable energy more broadly) in varying ways, raising interesting questions about the management of bilateral relations. Such more integrated analysis could also inform a further advance of the understanding of the changing geopolitics of energy that could benefit from combining consideration of a range of impacts and political responses. What is more, such more integrated analyses are likely to require regular updates and revisiting because the developments implied by the climate transition are, as mentioned above, set to evolve over time (with some candidates being the rise of circularity and of negative emission technologies).

Regular updates and reviews are also required because of unforeseen new developments, such as most recently the Covid-19 crisis. Although its wide-ranging effects are still further evolving, it seems clear that the pandemic has placed huge additional strains on national, European, and international governance structures and has generated enormous economic uncertainty. Lockdowns and restrictions on international travel have sharply reduced demand for fossil fuels in Europe and beyond, leading to unprecedented price shocks in international markets for oil, gas, and other commodities. The fallout of the crisis has provided further evidence of the vulnerability of fossil fuel exporters and has reinforced the urgency of economic diversification. At the same time, the recovery packages countries have designed have created the opportunity to accelerate the climate transition. Analytically, however, the crisis has reinforced the need to consider the decarbonisation challenge in an integrated way and in its broader context.

Fourth, there is a need and significant potential for an improved understanding of the interaction of different constellations of factors that shape the impacts of the climate transition on partner countries and bilateral relations. The present study has identified and taken into consideration a number of relevant factors, including state fragility, security context, and domestic climate policy. However, given the explorative character of our investigation, we have admittedly not been able to systematise particular constellations, also due to the large number of factors and their variation. A worthwhile step may therefore be to more systematically compare different cases and countries in order to identify varying combinations of factors that may produce or facilitate particular outcomes. This may also require a more in-depth conceptualisation and investigation of factors such as the regional security context. We hope that the results of our explorative study may constitute a sound basis for such more systematic and in-depth studies.

Fifth, there is significant scope to further refine and broaden the investigation and understanding of EU foreign policy. As mentioned in Chapter 1, our study has put the emphasis on the supranational EU level. We have taken into account the EU member states only at an aggregate and overall level, although – as also indicated – many if not most areas of external relations feature strong member state competences and actorness. Consequently, future research can further advance the understanding of options and strategies available to the EU (as comprising both the supranational and the member-state levels) by more systematically and comprehensively reviewing and taking into account the capacities and relations of the member states, starting with those that are prominent with respect to specific partner countries. In this context, Brexit may also prominently raise the issue of the scope and promise of coordination with non-EU member states such as the United Kingdom.

Sixth, the approach that we have just begun to develop here may obviously also be applied more broadly to other actors than the EU. Whereas our interest has been specifically in shedding light on EU external relations, there is nothing to suggest that similar questions could – and should – not also be asked and investigated with respect to the external relations of the United States, China, Japan, and others. The EU may still be at the forefront of the decarbonisation process by international comparison (Wurzel et al. 2017; Oberthür and Dupont 2021). However, the climate transition is eventually a global challenge and increasingly acknowledged as such, so that also the recalibration of external relations – and eventually of international relations at large – is a global challenge facing all countries individually and jointly.

Seventh, and consequently, a logical further step in the development of the research agenda may be the exploration of other implications beyond foreign policy and external relations. We have approached the challenge as one for individual actors (including composite/collective actors such as the EU) and their bilateral relations. However, these bilateral relations obviously form part of and shape broader contexts. As such, research may also usefully focus on and investigate how such broader contexts are challenged by, and can be fruitfully further developed, under European and global decarbonisation. Hence, future research may for example explore the challenges and opportunities arising from the climate transition for broader issues such as regional and international security contexts, regional and interregional relations, and international relations and the international order more broadly. We would hope that our investigation of (EU) external relations can form a steppingstone for such broader efforts.

The identified seven avenues for future research do not constitute an exclusive list, but they illustrate the potential and promise of research on (EU) external relations and foreign policy under decarbonisation. They form part of a rich research agenda that we hope the present study helps to move forward. While this study could naturally not answer all questions, we hope that the tentative answers it was able to suggest help substantiate the questions that need to be asked to align EU external relations with the policy imperatives of the global and European climate transition and to stabilise international affairs under these challenging and turbulent circumstances.

References

Adelle, C., Biedenkopf, K., and Torney, D. (eds.) (2018) *European Union external environmental policy: Rules, regulation and governance beyond borders*. Houndmills: Palgrave Macmillan.

Blondeel, M., Bradshaw, M.J., Bridge, G., and Kuzemko, C. (2021) 'The geopolitics of energy system transformation: A review'. *Geography Compass*, e12580. https://doi.org/10.1111/gec3.12580.

de Jong, S., Auping, W.L., Oosterveld, W., Usanov, A., Abdalla, M., Van de Bovenkamp, A., and Frattina della Frattina, C. (2017) *The Geopolitical Impact of Climate Mitigation Policies: How Hydrocarbon Exporting Rentier States and Developing Nations Can Prepare for a More Sustainable Future.* The Hague Centre for Strategic Studies. Available at: https://hcss.nl/report/the-geopolitical-impact-of-climate-mitigation-policies/ (accessed 28 June 2021).

EEAS – European External Action Service (2019) 'Colombia and the EU Hold a New Edition of Their High Level Political Dialogue'. Retrieved 26 July 2019 from https://eeas.europa.eu/headquarters/headquarters-Homepage/65502/colombia-and-eu-hold-new-edition-their-high-level-political-dialogue_en.

European Commission (2019) *The European Green Deal. 11.12.2019. COM(2019) 640 final*. Available at: https://ec.europa.eu/info/sites/default/files/european-green-deal-communication_en.pdf (accessed 20 June 2021).

European Commission (2021a) 'European Union. Trade in Goods with Nigeria'. Retrieved 07 July 2021 from https://webgate.ec.europa.eu/isdb_results/factsheets/country/details_nigeria_en.pdf.

European Commission (2021b) 'European Union. Trade in Goods with Indonesia'. Retrieved 14 July 2021 from https://webgate.ec.europa.eu/isdb_results/factsheets/country/details_indonesia_en.pdf.

European Commission (2021c) 'European Union, Trade in Goods with Azerbaijan'. Retrieved 13 July 2021 from https://webgate.ec.europa.eu/isdb_results/factsheets/country/details_azerbaijan_en.pdf.

European Commission (2021d) 'European Union Trade in Goods with Colombia'. Retrieved 13 July 2021 from https://webgate.ec.europa.eu/isdb_results/factsheets/country/details_colombia_en.pdf.

European Commission (2021e) 'European Union Trade in Goods with Qatar'. Retrieved 14 July 2021 from https://webgate.ec.europa.eu/isdb_results/factsheets/country/details_qatar_en.pdf.

European Commission (2021f) 'European Union, Trade in Goods with Canada'. Retrieved 14 July 2021 from https://webgate.ec.europa.eu/isdb_results/factsheets/country/details_canada_en.pdf.

Fund for Peace (2021) 'Fragile States Index 2021'. Retrieved 26 July 2021 from https://fragilestatesindex.org/.

Gemenne, F., Barnett, J., Adger, W.N., and Dabelko, G.D. (2014) 'Climate and security: Evidence, emerging risks, and a new agenda', *Climatic Change*, 123(1), pp. 1–9.

Gurol, J. and Starkmann, A. (2021) 'New partners for the planet? The European Union and China in international climate governance from a role-theoretical perspective', *JCMS: Journal of Common Market Studies*, 59(3), pp. 518–534. https://doi.org/10.1111/jcms.13098.

IRENA – International Renewable Energy Agency (2019) *A New World: The Geopolitics of the Energy Transformation*. Available at: http://geopoliticsofrenewables.org/Report (accessed 29 June 2021).

Kavvadia, H. (2021) 'The European investment bank's "quantum leap" to become the world's first international climate bank', *Politics and Governance*, 9(2), pp. 185–195. https://doi.org/10.17645/pag.v9i2.3921.

ND-GAIN – Notre Dame Global Adaptation Initiative (2020) 'ND-GAIN Country Index'. South Bend, IN: University of Notre Dame. Retrieved 05 July 2021 from https://gain.nd.edu/our-work/country-index/.

Oberthür, S. and Dupont, C. (2021) 'The European Union's international climate leadership: Towards a grand climate strategy?', *Journal of European Public Policy*, 28(7), pp. 1095–1114. https://doi.org/10.1080/13501763.2021.1918218.

Ohnesorge, L. and Rogge, E. (2021) 'Europe's green policy: Towards a climate neutral economy by way of investors' choice', *European Company Law*, 18(1), pp. 34–39. https://kluwerlawonline.com/JournalArticle/European+Company+Law/18.1/EUCL2021005.

Petri, F. and Biedenkopf, K. (2021) 'Weathering growing polarization? The European Parliament and EU foreign climate policy ambitions', *Journal of European Public Policy*, 28(7), pp. 1057–1075. https://doi.org/10.1080/13501763.2021.1918216.

Piggot, G., Verkuijl, C., van Asselt, H. & Lazarus, M. (2020) 'Curbing fossil fuel supply to achieve climate goals', *Climate Policy*, 20(8), pp. 881–887. https://doi.org/10.1080/14693062.2020.1804315.

Rayner, T. (2021) 'Keeping it in the ground? Assessing global governance for fossil-fuel supply reduction', *Earth System Governance*, 8, 100061. https://doi.org/10.1016/j.esg.2020.100061.

Swatuk, L.A., Thomas, B.K., Wirkus, L., Krampe, F., and da Silva, L.P.B. (2021) 'The "boomerang effect": Insights for improved climate action', *Climate and Development*, 13(1), pp. 61–67.

Torney, D. (2015) *European climate leadership in question: Policies toward China and India*. Cambridge/London: The MIT Press.

UNDP – United Nations Development Programme (2020) 'Human Development Report'. Available at: http://hdr.undp.org/en/countries (accessed 26 July 2021).

World Bank (2021a) 'Worldwide Governance Indicators 2019', Retrieved 26 July 2021 from http://info.worldbank.org/governance/wgi/#reports.

World Bank (2021b) *Regulatory Indicators for Sustainable Energy (RISE)*. Retrieved 26 July 2021 from https://rise.esmap.org.

Wurzel, R.K.W., Connelly, J., and Liefferink, D. (eds.) (2017) *The European Union in international climate change politics: Still taking a lead?* Abingdon: Routledge.

Index

Note: **Bold** page numbers refer to tables; *italic* page numbers refer to figures and page numbers followed by "n" denote endnotes.